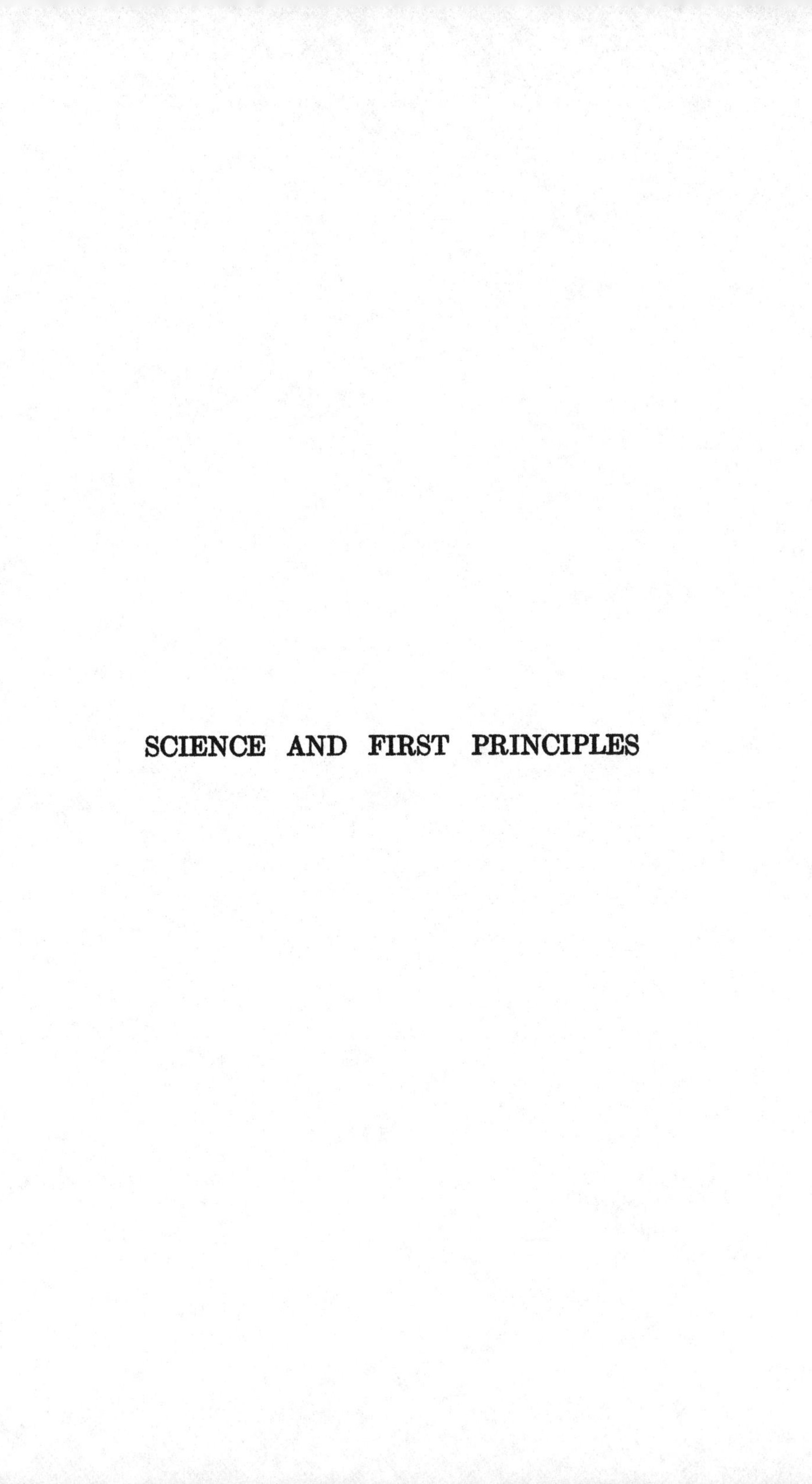

SCIENCE AND FIRST PRINCIPLES

SCIENCE
AND
FIRST PRINCIPLES

BY

F. S. C. NORTHROP

ASSOCIATE PROFESSOR OF PHILOSOPHY
YALE UNIVERSITY

CAMBRIDGE
AT THE UNIVERSITY PRESS
1931

CAMBRIDGE UNIVERSITY PRESS
Cambridge, New York, Melbourne, Madrid, Cape Town,
Singapore, São Paulo, Delhi, Mexico City

Cambridge University Press
The Edinburgh Building, Cambridge CB2 8RU, UK

Published in the United States of America by Cambridge University Press, New York

www.cambridge.org
Information on this title: www.cambridge.org/9781107695665

First published 1931
First paperback edition 2013

A catalogue record for this publication is available from the British Library

ISBN 978-1-107-69566-5 Paperback

To

C. J. N.

Deems Lectures Delivered in New York University in May 1929

PREFACE

Science proceeds in two opposite directions from its many technical discoveries. It moves forward with the aid of exact mathematical formulation to new applications, and backward with the aid of careful logical analysis to first principles. The fruit of the first movement is applied science, that of the second theoretical science. When this movement toward theoretical science is carried through for all branches of science we come to first principles and have philosophy. This book is a product of the last movement. Stated bluntly, it aims to determine precisely what contemporary scientific discoveries in many different branches of science reveal, and what all this means for philosophy.

The method which is used is quite simple. The first or philosophical concepts and principles involved in a given verified theory are those which it takes as primary. The primary concepts and principles of any theory are those which are taken as undefined, and are used to define other derived concepts and principles. Hence our task is the purely impersonal and objective one of dissecting the given scientific theories which our technical scientists have verified, to determine what concepts and principles are taken as primary or undefined.

We have attempted to do this with the major strategic theories of the major divisions of contemporary science. The question concerning whether a philosophy of contemporary science exists, is the purely objective query as to whether a common set of primary concepts appears, when such an extensive analysis is made. This is a question that cannot be resolved ahead of time, although the fact that all branches of science are dealing with the same universe indicates an affirmative answer. But even so, science may not have gone far enough in its investigations to determine what it is.

It happens that such analyses of current scientific theories give us a common set of primary concepts. However, it soon becomes evident that they are incomplete, for circular definitions appear, and one fundamental problem common to all branches of science exhibits itself. This indicates that the difficulties of current science center in first principles. Because the available common concepts necessitate certain consequences which have not been recognized, and because of certain facts which contemporary investigations have revealed, rather unequivocal clues to the nature of the missing principle are at hand. Thus our analyses lead us to the discovery of a new theory of the first principles of science. It appears that this theory, notwithstanding its simplicity, is sufficiently fertile to give rise to the confusing oppositions of current science and the many major aspects of human experience.

But three main points need to be grasped to sense the full force of our theory: first, the primacy of motion, and all that this entails; second, the necessity of defining the uniform, constant, structural, and continuous aspects of nature in physical terms; and third, the specific nature and status of the psychical. Once these three points are understood everything that we say follows.

The key to the second of these points was discovered by the writer while a graduate student at Harvard University, as a result of an analysis of the foundations of the type of biological organization revealed in Professor L. J. Henderson's nomogram of the blood. I am greatly indebted to him for placing his findings at my disposal at that early date. The results were presented in a crude form in my doctor's dissertation on The Problem of Organization in Biology, which is deposited in the Harvard University Library.

The essentials of the present book were given in six lectures delivered on the Deems Foundation at New York University in May 1929. I wish to express my appreciation to the Trustees of the Deems Lectureship for this honour and privilege. The lectures were delivered extemporaneously, and the manuscript has been completely rewritten since that time. But except for minor additions in the form of substantiating evidence and more exact and complete exposition, no change in essentials has occurred.

Certain conventions used in the text may be noted. The term atom is used in its original philosophical meaning, as the last indivisible substance. Hence it is to be identified with the electron rather than the chemical element in current scientific theory. For the same reason the one and only macroscopic atom is a simple substance not made up of parts. Reference books and the minimum essentials for a bibliography are designated at the end of each chapter. No attempt has been made to make this list exhaustive.

Throughout, the aim has been to attain technical scientific exactness and philosophical comprehensiveness with brevity. The latter quality is essential if the reader is to sense the common philosophical foundations of apparently diverse branches of science and experience. For this reason much can only be suggested or barely indicated. But if the reader takes the book as a whole, points merely suggested in one chapter may become reinforced elsewhere. Hence the reader is urged not to limit himself to the section of his special interest. In the long run this means a genuine economy of thought, as the Greeks perceived. They saw no point in solving a scientific problem twelve times over in twelve different branches of science, as if it were a unique and novel difficulty in each case, when it can be resolved once for all, with one-twelfth the effort, if attacked at the level of first principles. It appears that modern science can make more use, than it has, of this method.

I wish to thank the Cambridge University Press and the Yale University Press for permission to reproduce from the books of J. Barcroft, J. S. Haldane, and L. J. Henderson, the graphs that appear in Chapter IV, and the Editors of The Monist, The Journal of Philosophy, and the Proceedings of the National Academy of Science for permission to reprint from my articles in these journals, certain portions which appear in the last part of Chapter II. I am indebted to the Yale Graduate School for a grant from the Sterling Fellowship Fund, which was of great assistance in connection with the material of Chapter II; also to the Department of Philosophy and the Yale University Council's Committee on Research in the Social Sciences for stenographic aid in connection with the preparation of the manuscript.

Often I have had occasion to call upon members in other departments than my own for aid, particularly the Department of Zoölogy. Their suggestions have always been most valuable. I am especially indebted, however, for their interest and criticism, to my colleagues and present and former pupils of my own department, especially to Professor C. K. Davenport, now of the University of Virginia for suggestions that appear in Chapter V, and to my colleague, Dr. F. P. Hoskyn, for incisive comment which has improved many themes of this book.

My indebtedness to my teachers is greater than any acknowledgment here can express. In particular, the ideas of H. M. Sheffer, L. J. Henderson, and A. N. Whitehead have guided me. At times, I have been forced to dissent from certain views which Professor Whitehead holds, but this has happened because of an attempt to provide a solution for problems which he revealed to me. And even in dissent, I trust that I am following the true spirit, if not the exact letter, of his teaching. But in the end, I doubt if the differences are as great as at first appears. For the most part, I have but stated, in terms of the physical theory of nature, what he has uttered in terms of the functional theory. In any event there is no one with whom I would rather agree, and to whom I am more deeply indebted.

F. S. C. NORTHROP

New Haven,
November 29, 1930.

CONTENTS

SCIENCE AND FIRST PRINCIPLES

CHAPTER I

The Significance of Contemporary Scientific Thought

Certain things have happened in contemporary science which are of great importance. Only upon two other occasions in its twenty-seven centuries of history has the Western World faced a similar situation. In these instances every branch of human interest and endeavor was altered. There is every reason to believe that current developments will be no less far-reaching in their consequences. The theory of relativity, new conceptions of the basic nature and mechanism of matter, and new discoveries in biology have brought the first principles of science into question.

If the full significance of this fact is to be grasped, it is necessary that we view our present unusual situation in a background of scientific and philosophical thought which is very much wider than that of the last three centuries. Not only does the modern world fail to provide an analogy to our present predicament, but a knowledge of the ancient past is necessary for an understanding of current novelties. Conceptions have been rejected by Einstein and others which have been taken for granted in science and philosophy since the days of Leucippos, Plato, and Aristotle. A knowledge of the reasons for the original introduction of the rejected concepts is essential, if we are to understand our own time. This necessitates that we come to the meaning of contemporary science through the work of the ancient Greeks.

Our task is to determine what the important scientific conceptions of our day really mean. It is evident that something has happened which must eventually change the philosophy of each one of us. Nor must anyone suppose that the problem which we face affects only our philosophy; a solution of it is even more necessary for science. At the present time technical scientists find themselves with equations which they are unable to interpret, and with theories that they cannot make intelligible without statements of a philosophical character which certain of their colleagues do not accept. Stated bluntly, this means that scientists do not fully understand their own discoveries because the first principles, which make all technical discoveries intelligible, are in a state of flux. This is the reason why physicists, like Eddington, and Einstein, and Whitehead, and physiologists like Driesch, and Haldane, and Henderson, and pure mathematicians like Brouwer, and Hilbert, and Weyl, are writing philosophy. A new spirit and temper is abroad in the world. The old frame and background which contained the contents of the scientist's picture of nature, has dissolved in his hands, and he is being forced to shift his attention from the constituent details to the general structure, in order to prevent science from being overwhelmed by its discoveries, and destroyed by its very success. A change which strikes to the very foundation of things is upon us. The interesting and inevitable question arises concerning what this change is and what it entails. To this problem we are addressing ourselves.

Let us have no misgivings concerning the difficult character of such an undertaking. One simple fact will make this evident. The theory of relativity, to cite but a single instance, has led to different conceptions of the philosophy which it implies. If our analysis is to give rise to anything more than another opinion, it is necessary that we bring our conclusion to the test of fact. To determine the facts behind the first principles of science is an entirely different and more difficult task than is involved in the case of a specific technical scientific theory. In the latter instance, some crucial contemporary experiment will suffice; in the former, one can take nothing for granted. This means that one must go back behind all scientific theories to the facts which gave rise to them. Eddington's observation of the deflection of light rays passing the sun, may suffice to verify the theory of relativity, but once that theory has brought the first principles of science into question, a knowledge of the evidence behind traditional first principles is necessary as well. That evidence is to be found, only in the history of science, and, except for certain modifications added by Galilei and Newton, in the work of the ancient Greeks. Hence, it is essential that we come to a determination of current consequences through a study of ancient origins. When first principles are in question, one must begin at the beginning.

GREEK SCIENCE

The coming of science was heralded, some twenty-seven centuries ago, by a Greek named Thales, when he said, "All things are full of gods." This is not the purely pious remark which a first reading may suggest. Instead, it is a diplomatic way of saying that observed factors must determine one's conception of things. Before his time the universe had been conceived after the analogy of a glorified Punch and Judy show, in which gods and demons behind the scenes controlled one's body and the forces of one's environment in ways which man could neither understand nor control. Observed fact was fitted to unverified theory. The dictum of Thales changed this. It announced that the causes of things are in things themselves, where they can be studied and known; theory can and must be fitted to fact. Control of the minds of men by such a conception, made science inevitable.

The Physical Theory of Nature

When the Greeks proceeded from observed nature to scientific theory, two inescapable facts immediately impressed them. Thales[2] noted that nature is stuff, and

Heraclitos,[1] that it contains change. And what is more remarkable still, they had the genius to put these two facts upon record and force science to deal with them in its theory. Prosaic as it may sound, this is, nevertheless, the hall mark of science. The key to its secret is the discovery that nothing is too obvious to be important. In fact, one difference between science and religion centers right here. Religion tends to explain the obvious in terms of the complex and the mysterious; science, on the other hand, lays hold of the simple and obvious and by mastering it, beholds the miracle of the complex dissolving into clarity with the inroad of elemental principles. Baskets had balanced on the ends of wooden or iron rods in every market place in the civilized world since man first bought and bartered, but the science of statics did not arise until Archimedes put this obvious fact on record and used it to illuminate more complicated phenomena. Heavy objects had fallen to the ground since the earth first congealed into solid materials, but the mechanics of our universe was not understood until Galilei forced science to recognize it explicitly, and Newton traced the conception which it entailed to the foundation of the entire astronomical system. Similarly, the birth of one of the greatest scientific theories of all time was not far off, when Thales and Heraclitos noted the two extensive facts of stuff and change. Such is the importance of the obvious in the hands of the scientist.

Another characteristic of the observations of Thales and Heraclitos needs to be noted. They do not rest upon an accumulation of instances. The Greeks are beginning with observed nature as a whole, exactly as we begin, when we enter a room for the first time. General extensive characteristics are noted first. Hence the proposition that nature is stuff does not mean that we believe nature to be stuff because this book, this desk, that table, and that chair are physical. Instead, it is an immediately observed extensive fact which the whole object of observation exhibits to us. For this reason it has a universality which the findings of the technical sciences do not possess. This universality enabled the Greeks to state it in terms of a universal principle. Henceforth, we shall refer to it as the principle that the real is physical.

After Thales and Heraclitos had indicated that the real is physical, and changes, one of the greatest minds of all time appeared. It bore the name of Parmenides.[1] It was his genius to perceive that the presence of these two extensive characteristics in the same universe constitutes a problem. He made this clear by demonstrating that they contradict each other. The procedure which he used in establishing this is interesting and important. He first specified what the two facts of stuff and change involve. Then he proved that the existence of one is incompatible with that of the other, if nothing else is assumed. It immediately followed that something more must be assumed. Moreover, this additional factor was such that it could not be observed. Thus, the two obvious facts of stuff and change led men to the knowledge of factors in nature which cannot be grasped completely by the senses. In this fashion, man discovered that nature involves more than appears on the surface of things.

There is nothing unusual about such a state of affairs. It happens again and again in the history of science. Observed or verified principles are discovered which contradict each other, and new principles have to be introduced to escape the contradiction. We shall find precisely the same thing giving rise to the discovery of the special theory of relativity with Einstein. This point is important because it throws light on the power of science. It is said by uninformed minds again and again that certain factors of human experience are beyond the reach of science because they are not immediately observable; the implication being that science must limit itself to the purely observable. Nothing is further from the truth. Science is dealing with the unobservable every day of its existence. It has done so from the very beginning of its history, and it is doing so at the present moment. It is because of this, that we can be certain of nature being different from what it appears to be. The distinction between the world known by sensation or observation and the world known by reason arises. But this does not mean that scientists are speculating or fitting facts to theories. The observed facts themselves must determine whether such a distinction must be drawn. Parmenides' procedure merits our attention because it provides the criterion. Science begins with fact, using the method of observation. If the deliverances of observation gave rise to propositions which were compatible with each other, science would continue to conceive of nature exactly as it appears to be. It happens, however, as we have just indicated, that observation gives rise, in many cases, to undeniable propositions which contradict each other. When these contradictory propositions can hold for the same universe only if nature is conceived as other than it appears to be, then the distinction between appearance and reality becomes an accepted doctrine of science.

Note how this occurred in Greek science. Parmenides first specified what the fact of stuff entails. He indicated that it involves, not merely the proposition that the real is physical, but also the principle that real is being, where being means that the real does not change its properties. In other words stuff means permanence.

The evident and rather abstract character of the principle of being must not cause us to overlook its importance. In fact, the rôle which it plays in science is surprising to those who are not accustomed to think about science from the point of view of first principles. In the first place, it is the basis of the doctrine of mechanical causation. We have heard a great deal about this doctrine in the modern world. The unassuming little principle of being is its real justification. For the principle of being means that the real causes of natural phenomena do not change their properties. Hence if they are discovered once they are determined for eternity; the passage of time will introduce nothing fundamentally new in the realm of causes.

This thesis is the doctrine of mechanical causation.[16] Secondly, the principle of being makes science possible, because it justifies the logical principle of identity. Unless meanings and ideas remain fixed, we cannot think, and unless this fixity applies to nature in some fundamental and approximately universal sense, thought cannot apply to nature and science is out of the question. This fixity is also necessary to make valid inference and prediction possible. These considerations suffice to indicate that one cannot play with the first principles of science without running the risk of destroying science itself. We shall find that this actually happened at one stage in its history.

Finally, the principle of being means that the notion of eternity is more fundamental than the idea of temporality. The importance of this point cannot be overestimated. We shall find ourselves saying a great deal more about it when we consider twentieth century science. The idea that reality is eternally what it is, is a necessary part of our observation of the extensive fact of stuff. Had this idea been fully grasped we should not have fallen into certain of the errors which Einstein has had to correct. For the priority of eternity means that we do not come to nature perceiving it at an instant in an infinite time series; we observe it as something which is eternal first, and come upon the discovery of temporality in its parts later. Once this idea is grasped, it does not surprise us that time is relative. This point is important also as a justification of scientific induction, since it means that the observation of the fact of stuff escapes the method of accumulation of instances in a temporal as well as a spatial sense. We do not observe nature to be stuff at an instant, and then add up all the instants in our short lives to infer that it will be stuff always. Such a notion presupposes an unverified complicated psychological and physical theory of time for which neither immediate experience nor recent physics gives any justification. We observe nature to be extensive stuff which involves permanence as a part of its very nature, and the notion of time is a local detail. It is to be noted that the extensive fact of stuff can remain true, notwithstanding the change of individual physical objects. Such is the significance of a science which is reared on the extensive characteristics of observed nature as a whole. It was an event of no mean significance when Thales and Heraclitos observed the two extensive facts of stuff and change, and Parmenides noted that the fact of stuff involves the principle that the real is being as well as the principle that the real is physical.

Once this was recognized Parmenides had no difficulty in proving that the two facts of stuff and change contradict each other, if nothing more is assumed. The proof is absolutely sound; and so brilliant in character as to be almost humorous. Change, he said, must be due to generation or to motion. It cannot be due to generation for that means that the real changes its properties, and is incompatible with the principle of being which stuff entails. But neither can it be due to motion, if stuff is conceived as nothing but one physical substance or many microscopic particles. For motion requires that a thing moves from where it is to where it is not. If nature is nothing but the stuff which moves, there is no 'where-it-is-not', and hence motion is impossible. The difficulty is not escaped by regarding stuff as many, rather than one. For the motion of many particles involves a 'where-it-is-not' as much as the motion of one; a difficulty is not met by multiplying it many times. Moreover, there cannot be many particles if nothing but the stuff of the moving particles is supposed to exist. For manyness requires something to enable one to distinguish between one atom of stuff and another, and this is impossible if nothing but the stuff of the atoms exists. The essential point in the latter argument is not so much the need for an intervening space, as the necessity of something to designate the difference between one particle and another. In a kinetic atomic theory the category of stuff gives only the respect in which the atoms are identical or one; it cannot prescribe the respect in which one atom is other than another. Stated positively this means that one atom can be distinguished from another only in terms of its unique relation to some common referent. If nothing but the stuff of the microscopic particles exists there is no such referent. Hence atomism is impossible.

It is to be noted that this argument takes care of those who would attempt to define atomic motion in terms of the relation of the microscopic atoms to each other. Before there can be even relative motion, there must be many atoms, and the manyness, wholly apart from their motion, is meaningless and impossible, unless there is a common referent other than those atoms.

Heraclitos and Parmenides attempted to escape this contradiction between the two extensive facts of stuff and change by denying one of the facts. Thus the former maintained that the real is a flux in which matter is a mere appearance which comes and goes, whereas Parmenides asserted that nature is one huge solid spherical substance and that change is an illusion. But the Greeks' sound scientific sense prevented such a patent denial of fact. Observation revealed that neither of these two obvious extensive characteristics of nature can be dismissed in this facile fashion; a theory was demanded which could admit both factors as real, and do justice to Parmenides' logic also. Since Parmenides had proved that change cannot be due to the motion of changeless particles of stuff if there is nothing but these particles, it followed necessarily as Leucippos noted,[1] that a referent in addition to the microscopic atoms must exist. When this referent was identified with the spatial characteristic of nature, the doctrine of absolute space came into scientific theory, and the kinetic atomic theory was put in the form which it has maintained down through modern times.

Note how this conception meets the difficulties to which Parmenides' analysis gave expression. The physical character of the atoms accounts for the extensive fact of stuff; their motion for the extensive fact of change. Since both the atoms and space possess fixed properties, the principle of being is satisfied. Since space exists independently of the microscopic particles, a meaning exists for the place where they are not, which motion requires, and a basis is present for distinguishing between one atom and any other, in terms of its unique relation to this common referent.

It has been supposed that this placed the physical theory of nature upon sound foundations. Zeno proved that this is not the case. He showed that if the referent for motion is continuous space, then the movement of a body through a finite distance in a finite time should be an impossibility. The proof is simple and valid. In a continuous space there are an infinite number of points in any finite distance. Hence, to move through a finite distance, means, if motion is in space, that a body must pass from one point to another an infinite number of times. This is impossible in a finite period of time.[1] Obviously, the Greeks should have concluded that a referent other than the microscopic particles exists, and that it is not absolute space. Instead, science took over the physical theory of nature as Leucippos, who was a successor of Parmenides, stated it.

Certain consequences of this philosophy are of interest. In the first place, all relations, except spatial ones, are varying effects of atomic motion. Order or form is never a cause of natural phenomena; it is the mere casual effect of a motion which might produce disorder as easily as order. In fact, chaos must win out over organization in the long run, as modern science later asserted in the second law of thermodynamics. This accidental character of relations is one of the sources of the modern conflict between science and religion. Because of it, the order of nature can never be taken as a proof of design. Secondly, nature is different in detail from what it appears to be. The gross motion and stuff is real, but it is a compound of unobservable moving microscopic parts. Thus, the physical theory of nature gives rise to an epistemological distinction between the world known by reason to which scientific theory refers, and the world of sensation which observation reveals. Nevertheless, there is no gulf between them. The microscopic particles are the gross objects and motions which we perceive. Finally, this philosophy of science is connected to the facts upon which it rests by the necessary relation of logical implication. One cannot admit the two extensive facts of stuff and change and fail to accept the kinetic atomic theory without involving oneself in a contradiction. Since its facts are obvious extensive characteristics of observed nature, and possess a unversality which findings of the technical sciences cannot enjoy, we must expect this theory to possess a certainty and a lasting quality which none of its rivals can equal. History reveals this to be the case.

Such is the fertility of the obvious in the hands of men who can state its deliverances in explicitly formulated propositions and follow those propositions to their consequences. It was not a trivial event when Thales and Heraclitos forced science to face the two extensive facts of stuff and change. Nor was it an evidence of unwarranted fussiness when Parmenides insisted upon stating those facts in formal terms, and deducing their consequences. Thereby the power of logic when combined with empiricism was revealed, and the foundations were laid for the discovery of the only scientific theory which has the distinction of lasting from the very dawn of scientific endeavour to the present moment. Henceforth, we shall refer to it as the physical theory of nature.

The Mathematical Theory of Nature

The second major movement in Greek science occurred in mathematics and astronomy.[4] In the modern world physics has been the dominant science. It must never be forgotten that mathematics occupied this privileged position in the Greek period.

One of the first Greeks to appreciate the significance of mathematics for an understanding of nature was Thales.

He is reported to have imported from Egypt that piecemeal type of information which the Egyptians and Babylonians had gathered. One of the most important advances came with a follower of Thales named Anaximander.[1] He was impressed by the unlimited continuity which physical nature exhibits. This extension and continuity of the physical he designated by the doctrine that the real is the "Boundless". This "Boundless" he regarded as physical.[2]

This continuous character of nature presented a problem: If the universe is physical and is a continuum, why do the local discontinuities and differentions in it occur? To this the atomic theory finally gave a partial answer. Before Leucippos, the Pythagoreans attempted a different solution.

A study of living organisms and music suggested to Pythagoras[1] that things in nature are a balance between opposing factors. By combining this idea with the physical "Boundless" of Anaximander, he was led to the conclusion that there must be a "Limit" standing over against the "Unlimited" or "Boundless" to provide the necessary second pole of the equilibrium. This "Limit", which was unknown, was supposed to restrict the unlimited possibilities of the "Boundless" to the specific differentiations which actually exist.

This theory had one weakness. The "Limit" was unknown and its attributes were unspecified. Had this been the end of Pythagorean thought, science would have stagnated in its midst. For investigation could always have been avoided by referring the existence of anything which was not understood to the action of the "Limit".

However, Pythagoras made a rather startling discovery which threw an entirely new light upon things. He noted in connection with music, that it is possible to express the specific nature of the particular equilibrium which a particular tone exhibits, in terms of arithmetical conceptions, without any reference to either the "Limit" or the "Unlimited". Generalizing this idea for the whole of nature, it suddenly dawned upon the Pythagorean mind that this universe is in some fundamental sense essentially numerical and mathematical. Pythagoras gave utterance to this conception in the doctrine that the real is number.[3]

An attempt was made to define number and geometry in terms of the discontinuous "pebbles" of the physical theory. This attempt broke down, for the Pythagoreans, with the discovery of the irrational. The mind of man was thus prepared for a new theory of nature.

The realization that mathematics holds the key to the structure and essential character of specific things, made this science of tremendous importance. It was natural that it should receive most serious study.

In this, Pythagoras led the way. Besides formulating and perhaps proving the proposition concerning the relation of the length of the hypotenuse of a right-angled triangle to the length of the other two sides, he analyzed the notion of number, and laid the foundation of number theory.[4] The latter branch of the science was carried on by Theaetetus and Plato. With the passing of time many isolated geometrical principles were being discovered. Hippocrates of Chios was the first to attempt a systemization of them. Gradually the accumulation of concepts and laws went on until the whole of geometrical science was thrown into a unified deductive form by Euclid.[6]

While these developments were taking place in mathematics, empirical evidence from an observation of the heavenly bodies was accumulating in astronomy. Slowly, this material began to fall into order, until Eudoxos finally organized the entire astronomical field to give the Western World its first systematic mathematical astronomy.[7]

In this great achievement a remarkable fact appeared. It must have come upon the realistic Greek mind with a distinct shock. The laws which brought several centuries of astronomical evidence into order, with a degree of precision which made accurate prediction possible, said not one word about physical objects. They referred instead to perfect geometrical forms, which Eudoxos warned his contemporaries against regarding as physical, and to ideal arithmetical proportions. Thus the ideal purely conceptual categories of mathematics and logic were revealed as constituting the very essence of the entire astronomical universe in which we live. Is it any wonder that the leading scientists of the Greek world concluded that the real is rational rather than physical? Had not nature revealed itself to them as a system of logical or mathematical forms rather than as a collection of moving physical atoms?

It was natural that such a scientific outlook, which had behind it several centuries of inductive investigation in the oldest and most respected of the sciences, should be given articulate expression in the form of a philosophy. It was inevitable also that the men who should do this, should be mathematicians and that they should note the trend and foresee the outcome even before the astronomy of Eudoxos was completed. It happens that their names are Pythagoras and Plato,[8] and that the title which this second major theory of first principles has born in history is Platonism.

At last we are able to understand why the Platonic philosophy affected Western civilization so much more completely than has been the case with the professional modern systems of philosophy. It had behind it the carefully analyzed conceptions of pure mathematics and the verified evidence of the science of astronomy, rather than mere epistemological opinion. We can understand also why Plato left his Academy to mathematicians, and why he placed over the door of that school in philosophy the words: "Only mathematicians need enter here." He was merely saying that no one need hope to master the first principles of his philosophy unless one is acquainted with the fundamental conceptions of the leading science of the day.

Once one appreciates the scientific basis and background

of the Platonic philosophy its essential principles become intelligible. Its thesis that the real is being and is rational is the essential doctrine of any period of scientific endeavor which regards mathematical categories as more fundamental than physical ones. The validity of the principle of being in the mathematical as well as the physical theory is evident, for mathematical forms are exhausted in their meanings, and these are eternal and immutable things. It is evident, therefore, that the mathematical theory of nature must be as mechanical as the physical theory, since the doctrine of mechanical causation is a necessary consequence of the principle of being. This, the mathematical astronomy of Eudoxos and Hipparchos clearly indicates.

The difference between the mathematical and physical theories of nature centers in the principle that the real is rational. This replaces the thesis that the real is physical, which is the distinguishing mark of the physical theory. The principle that the real is rational must not be misunderstood. It means very much more than that an intelligible account of natural processes can be given. In addition, it signifies that when such an account is gained, nature will be found to be made of ideal rational forms which only reason can grasp, rather than physical objects which can be observed or imagined. A mathematical equation comes nearer to the nature of reality than a physical atom.

There are three consequences of the mathematical theory of nature which demand attention. The first is the primary causal importance which it gives to relations. According to it, logic structure is that which science finds to be ultimate in nature. Thus, the significant rôle which mathematics plays in expressing its objective laws is made intelligible. A second inevitable consequence of the mathematical theory is epistemological in character. Plato stated it when he drew a distinction between the apparent world of sensation given in observation, and the real world of mathematical forms which is known only by reason. No one who accepts the mathematical theory of nature can escape this epistemological principle. Certainly nature appears to be physical and changing; this mathematical theory maintains, however, that it is essentially logical and changeless. Obviously, such a theory can be maintained before such facts only by holding that the real world of first causes is different in character from what nature appears to be. This point is very important. We shall find, when we trace the fate of this theory in history, that it led to the degeneration of Platonism and the death of science.

The third consequence of the mathematical theory is methodological in character. Since mathematical forms are not observed in nature and, as Plato said, are suggested by, but not contained in the world of observation, it follows that one cannot proceed, as did the physical theory of nature, from the facts of observation to one's scientific principles by the necessary relation of formal implication. The facts merely suggest the mathematical forms; they do not imply or contain them.[10] Hence, as Plato maintained, the fundamental scientific method in this theory is the method of hypothesis. Since this method always commits the logical fallacy of affirming the consequent, Plato tried to supplement it as much as he could. This led him to introduce the dialectic, which is not the vicious thing its modern connotation suggests to certain minds, but the simple sound idea that all hypotheses must be traced to their common presuppositions and unified into a consistent deductive system. When we attempt to do this for psychology and epistemology as well as mathematics and astronomy, we find ourselves face to face with all the problems with which Plato wrestled in his Dialogues. To see the second movement of Greek science in the light of its bearing on first principles is to understand the philosophy of Plato. It was the mathematical and rational character of the inorganic universe that made man an idealist in ancient times.

The Functional Theory of Nature

The third major movement in the Greek period involved the sciences of medicine and biology. Three men stand out in connection with it. They are Hippocrates of Cos,[11] Empedocles, and Aristotle.

Hippocrates was the first to emphasize that a living organism is a mechanical system. When man believes that disease is due to chance, thought is not given to its study, and medicine remains in the hands of quacks and soothsayers. It was an important event, therefore, when Hippocrates wrote that chance is but another name for ignorance, and urged his disciples to watch the course of disease and look for regular connections and causes. The science of medicine dates from the day when he called man's attention to the mechanical character of living processes.

In addition, he noted that one of the most essential characteristics of a living thing is its organization. The relation which joins the materials seems to involve something more than the properties of the materials themselves.

It happens that these two, mechanical and organic characteristics of life, are not easy to reconcile. In fact, they constitute a difficulty, the problem of organization, which concerned Aristotle, and which has baffled biologists to the present day. It will concern us when we come to contemporary developments. Hippocrates of Cos made no attempt to resolve it. He was content merely to indicate the facts.

It remained for Empedocles,[12] who had helped to develop the physicial theory of nature, to give the Western world its first theory of the organism. This happened when he attempted to account for birth, death, and structure in terms of the principles of the atomic theory. Notwithstanding certain somewhat fantastic conceptions, this mechanical theory of life was sufficiently fruitful to enable him to discover the principles of the struggle for existence and the survival of the fittest which were later to contribute to the fame of a great Modern. This theory, however, was doomed to lie dormant until our era, because the immature state of physics and chemistry in Empedocles' day made it impossible for any one to bring the atoms of his theory to bear in an effective way upon the specific processes of living things. This remained for the modern science of physiological chemistry.

At this point, Aristotle [14] came upon the scene. He was the son of a physician, and had spent twenty years amid the mathematical and astronomical ideas of the academy of Plato. Upon the death of his great teacher, he returned to the bays and islands of the Greek coast to spend several years observing living organisms. Out of these observations came the "Historia Animalium", one of the most truly inductive biological treatises ever written. Every student of Aristotle should begin with this work and, after completing it, proceed through the "De Partibus Animalium", the "De Incessu", the "De Anima", and the "Physics", to the "Metaphysics".[13] When this is done one will observe how the characteristics of living things led Aristotle to reject the philosophy of the ruling science of his day, which he had learned in the Academy, and rear science upon entirely new foundations.

The first fact to stand out amid all the differences and details which his tremendous mind had grasped, was generation. Living things grow and reproduce themselves. There is hardly a word which appears more often in his scientific treatises. The immature state of physics and chemistry made it impossible for him to reduce generation to the motion of atomic entities which do not change their properties. Greek biology would never have progressed beyond the work of Empedocles had he done this. No alternative remained, therefore, for one who would not fit biological facts into a preconceived theory, brought in from inductive natural philosophy, but to regard generation as ultimate.

Since irreducible generation means that the real changes

its properties, the acceptance of the principle of becoming was inevitable; the principle of being, and with it, the physical and mathematical theories of nature, had to go. The growth and reproduction of living things seemed to indicate that reality is a process.

The second fact to impress Aristotle was organization. It led him to precisely the same conclusion. For it suggests that a living thing involves not only the stuff of which it is made, but the form which that stuff exhibits. This was a new doctrine in Greek science, for it meant that both matter and form are causes. Once this was admitted both the mathematical and the physical theories had to be rejected, since the former admits only the existence of a formal cause, and the latter, only a material cause.

But this is not all. And here, we come upon a point which we must learn from Aristotle if we are not to fall into grievous error in our own reflections upon contemporary developments. He perceived that it is impossible to maintain both matter and form as causes, by merely retaining the physical atomic theory and adding on an organic relation as an additional causal factor. The idea of a disembodied form organizing physical atoms which cannot organize themselves is meaningless. Aristotle said precisely this when he emphasized that there is no form apart from something which has form.

This point can be made evident if we state it in modern terms. The principle of parsimony dictates that there is no justification for introducing a formal cause in addition to the material cause, unless a living organism exhibits an organization which the physico-chemical elements alone cannot produce. It follows, therefore, if a formal cause is necessary, that it must change the direction of motion of physical particles. Before such a task, however, a disembodied form is useless; only an external physical force will suffice. It becomes evident, therefore, either that all relatedness in nature reduces to physical causes, or that the physical theory of nature must be rejected.

The attempt of certain modern scientists to supplement

the physico-chemical categories with organic relations, or emergent evolution, or mathematical equations, when faced with a structure that presents difficulties for the physical theory can produce nothing but nonsense. Such a procedure is precisely as ridiculous as that of a pure mathematician who would expect to derail the Twentieth Century Limited, by attempting to think the equation for a disembodied open switch across its pathway.

Perceiving the impossibility of coupling the doctrine of a formal cause with the physical theory of nature, what was Aristotle to do? The fact of organization indicated the presence of both a material and a formal cause. Obviously, the physical theory had to be rejected. Matter, instead of being conceived as a substance had to be regarded as the attribute, along with form, of a more fundamental and essentially different type of substance.

Note how this meets the difficulty. Since both matter and form are essential attributes of a common substance, the irreducibility of one to the other, which biological organization seemed to Aristotle to involve, is intelligible. Moreover, since every substance always exhibits both its formal and its material attribute, the thesis that there is no such thing as form apart from something which has form, is justified. Furthermore, if matter and form are the mere passive attributes of a common substance, then the intermixture and interaction of them becomes explicable without any fallacy. This means, however, that the new more fundamental type of substance must be conceived as a process or activity, for only as it expresses its own nature in a single synthesis can the merging of its material and formal attributes take place. To say, however, that this new type of substance is a process or activity is to maintain that it changes its properties. Thus it happened that the fact of organization as well as the fact of generation drove Aristotle to the principle of becoming. Only by regarding nature as at bottom a process in which its two major attributes are synthesized can the doctrine of a formal cause and a material cause be made intelligible.

When Aristotle considered the order of inorganic nature which his colleagues, the astronomers, had revealed, the form which it involved, forced him to regard nature as one rather than many. Otherwise, the form, which joins organisms and stones and planets together to constitute the order of nature as a whole, becomes something existing independently of the things joined and the fallacy of a disembodied form recurs. Aristotle's sense of the reality of specific individual things, which his biological investigations had made so vivid, caused him to waver, and perhaps even desert the dictates of logic, on this point. This may well be one source of his designation of the "Unmoved Mover" as pure form. However, the development of this thesis must wait for another occasion. The important point to note is that the existence of form in the whole of nature, which does not reduce to physical causes, cannot be admitted, without regarding nature as one substance and rejecting all genuine atomism and individuation.

Moreover, to say that matter and form are attributes of a more fundamental type of substance is to maintain that they are not real or primary causes. For certainly passive attributes cannot cause anything. Aristotle said precisely this when he identified the substance and its activity with the efficient cause, and made the latter more fundamental than the material and formal causes.

One more step had to be taken before the doctrine of the irreducibility of generation and organization had revealed its last consequence. The reasons for the acceptance of the principle of becoming have been noted. It means that the real changes its properties. This means that new properties will appear in the efficient cause with time. Hence, the doctrine of mechanical causation must be rejected. Future effects cannot be predicted upon the basis of a knowledge of present factors. Final as well as present and past conditions must be considered. This is the principle of teleology.

This makes it evident that Aristotle's complicated doctrine of the four causes is not born out of the supposedly perverse habit of a philosophical mind to make unnecessary distinctions, but is, instead, an inevitable consequence of any scientific evidence which forces one to regard both matter and form as irreducible causes. For the latter conception has no meaning unless the principle of becoming is accepted and this principle in turn implies the principle of teleology.

To begin with Aristotle's biological treatises and to come to grips with the two facts of organization and generation which circumstances forced him to regard as ultimate, is to find oneself driven to the theory of the first principles of science which his philosophy represents.

It happens, therefore, that an essentially original philosophy arose when Aristotle turned aside from the mathematics and astronomy of the Academy to give his attention to the parts and habits and species of living things. Since it is uniquely designated by the principle that the real is becoming, we shall henceforth refer to it as the functional theory of nature.

Two consequences of this theory remain to be noted. The conception of nature as a dynamic unity, of which all factors are essentially inseparable attributes means that the analyses of science always involve a certain over-simplification and falsification. Hence, the most sound method of scientific procedure, aside from pure description, is the method of abstraction. One never passes from the observable world to a more real world which more accurately represents the nature of things. One merely centers one's attention upon particular aspects of the totality of fact in order to appreciate in more detail the unity which observation reveals. This comes out explicitly in current modern examples of the functional theory in which the method of abstraction receives great emphasis.[15] It was involved implicitly, however, in Aristotle's doctrine of the four causes. Reality is the synthesis of the four aspects, expressed finally as entelechy.

This methodological consequence of the functional theory entails an epistemological principle. The mathematical theory of nature maintains that the real is suggested by, but not given in, sensation. Aristotelianism holds, on the contrary, since the categories of science are but abstractions from observed nature, that the world of sensation gives real knowledge. This point happens to be very important. We shall find that it saved science after the Dark Ages.

The Problem of Matter and Form

The program which Thales initiated brought forth results. Special sciences arose in which three major scientific movements occurred. The first in inductive natural philosophy necessitated the conclusion that the universe in which we live is a system of changeless microscopic particles of stuff moving in absolute space; the second, in mathematics and astronomy, suggested that it is a system of ideal unobservable geometrical forms which only the reason can grasp; and the third in medicine and biology, led Aristotle to the conclusion that reality is a dynamic teleological process which exhibits matter and form as major attributes. Thus philosophy was born of science.

It is because Greek science failed to agree upon a single theory of first principles that it has such an obvious philosophical character. This must always be kept in mind during any comparison with modern developments. Modern science agreed on first principles. What is constant can be neglected. When agreement exists discussion is unnecessary. This made modern science apparently independent of philosophy. However, empirical evidence made this impossible for the Greeks. Hence, for them, to be scientific was to be philosophical. There was no escape from a consideration of first principles.

Moreover, it appears that we must revise our conception of philosophy. Moderns have been reading the modern departmentalization of knowledge back into the Greek period. This has led to the conception, conveyed by our text-books, that Greek philosophy is a single isolated movement of speculative thought which began with Thales and ended with Aristotle. Nothing is further from the truth. It is a product of three separate movements in Greek empirical science. To forget this is to divorce the conceptions of Leucippos, Plato, and Aristotle from the inductive evidence which gives them their validity, and to debase philosophy by reducing the fundamental questions of existence to a mere matter of individual taste.

The time has come for us to take the great theories of the Greeks off the dusty shelves of the Museum of Philosophical Systems upon which the modern world has placed them, and return them to technical science and nature which is the source of their vitality. When this is done the inseparable connection between philosophy and science will be rediscovered and philosophical conceptions will take on some of the certainty which scientific theories have enjoyed. It is because Greek philosophy had this character, due to its natural evolution from science, that it commands our attention and has established certain conclusions which are as certain today as when they were first discovered.

Let us summarize these certainties. They provide necessary materials for the solution of our contemporary scientific and philosophical problems. The three conceptions of nature which Greek science formulated are uniquely defined in terms of four basic principles: (1) the principle that the real is physical, (2) the principle that the real is rational, (3) the principle that the real is being, and (4) the principle that the real is becoming. The principle of being means that the real does not change its properties; the principle of becoming, that it does. To assert that the real is being and is physical is to maintain the physical theory of nature; that the real is being and is rational, the mathematical theory; and that the real is becoming, the functional theory.

The physical theory involves the following corollaries: (1) the principle of identity, (2) the priority of eternity over temporality, (3) the principle of mechanical causation, (4) the doctrine that all non-spatial relations are effects of atomic motion, and (5) the existence of a referent in addition to the microscopic particles.

The mathematical theory implies (1) the principle of identity, (2) the principle of mechanical causation, (3) the doctrine that only relations are causes, (4) the thesis that the method of hypothesis is the fundamental scientific method, and (5) the epistemological principle that the real world is suggested by, but not contained in, the world of sensation.

The functional theory involves (1) the principle of teleology, (2) the primacy of the method of abstraction, (3) the epistemological principle that the real world is contained in the world of sensation, (4) the doctrine that matter and form are mere attributes of a process, or "event", or dynamic type of substance, and (5) the thesis that there is only one real individual in nature which is the "event", or process, or dynamic substance, termed nature as a whole.[16]

What does Greek analysis and evidence indicate with reference to the respective merits of these three conceptions? The validity of any theory depends upon the universality of its facts and the necessity of the relations which joins it to these facts. Judged from this point of view it must be maintained, in opposition to accepted philosophical opinion, that the physical theory of nature rests upon the soundest ground.

It is the only theory which can lay any claim to the universal applicability of its premises. The extensive facts of stuff and change not only apply to the whole of observed nature but they have a priority and obviousness which necessitates that they must be either explicitly or implicitly involved in all technical investigations. The mathematical and functional theories arose in the special sciences. Notwithstanding the genius of their founders, they represent the type of philosophy which technical scientists produce, i.e., they are a suggested product of isolated local facts which are always a function of the emphasis of the moment. It will not surprise us, therefore, if Platonism and Aristotelianism pass out of existence, as seriously considered scientific conceptions, when new facts in new fields of specialized investigation appear, and technical science takes on a new interest. Nor will it appear as a miracle if the physical theory persists through all the new discoveries which time and effort may reveal. Certainly, it is difficult to imagine how the inevitable consequences of obvious extensive facts can be escaped by any age that is truly scientific.

The physical theory enjoys the same superiority with reference to the relation which joins its principles to the facts which would justify them. Of the three theories of nature, it is the only one, in which facts are connected to theory by the necessary relation of formal implication. According to the mathematical theory, as Plato pointed out, observed facts merely suggest theories, they do not imply them. The relation is not one of necessity. In the functional theory also, a variable psychological factor enters, since its fundamental method is the method of abstraction. Hence, the same facts may suggest different mathematical forms or different abstractions to different observers. However, once one has accepted the facts of the physical theory there can be no choice about accepting the principles of that theory, for Parmenides has proved that this cannot be avoided without involving one's self in a contradiction.

However, the differences between these three theories must not cause us to overlook what they have in common. They are the three answers and, since they exhaust the logical possibilities, they are the only possible answers to one fundamental problem. This problem must be stated somewhat abstractly. It is the problem of the relation between things and their relations, or in Greek terms, the problem of matter and form. The physical theory maintains that all relations except spatial ones, reduce to, or are effects of, the properties of matter. The mathematical theory affirms that nature is to be understood, and all science stated, in terms of relations. Whereas the functional theory takes the middle compromise position, maintaining that both things and relations are causes.

Since these three answers to this question exhaust the logical possibilities, it follows, unless science passes to some more fundamental issue, that the history of the first principles of science can be nothing more than the story of the fate of these three theories before historical circumstance and new evidence. History indicates that this is the case.

FROM THE GREEKS TO NEWTON

Following the Golden Age of Greek civilization the mathematical theory of nature came into the ascendency. This is easy to understand. It had behind it the authority which age and the accumulation of empirical evidence over several centuries had given to the sciences of mathematics and astronomy. It is to be remembered that these sciences carried the same conviction, for the Greeks, that physics has with us. Moreover, Aristotle was one of the last of the great Greeks. Hence his science of biology, and the functional theory of nature which grew out of it, could scarcely be expected to overwhelm the older rivals. In fact, the only serious opposition which Platonism received came from the physical theory of nature. But an inadequate physics and chemistry necessitated that its day of triumph was not at hand.

The Mathematical Theory of Nature and The Decline of Science

Gradually science proceeded into decline. There are good reasons for believing that the dominance of the mathematical philosophy was responsible for this. For this theory, with its doctrine of objective and unobservable ideal forms, does not become reasonable unless one accepts the conclusion that the real world is not what nature appears to be. But to accept this doctrine is to deprecate the importance of the observable world of sensation. A loss of interest in nature for its own sake follows. Without this interest, empirical science dies, and without empirical science, there is no mathematics or mathematical theory.

Furthermore, the emphasis upon the adequacy of purely mathematical formulations in science, leads not merely to a loss of interest in this world, but to an absorbing interest in some other one. Man centers his attention upon the forms known only by reason and neglects the world of sensation which suggests them. Guided by the scientific dogma that nothing is truly scientific which is not mathematical, man begins to regard the conceptual theories as more important, and possessing a greater authority, than the observable factors which provide their justification. Thus Plato is followed by Hypatia,[17] facts are fitted to theories, and science commits suicide. In this fashion, the gods are taken out of things, to be put back behind them again, and the work of Thales is undone.

Sooner or later man begins to reason as follows: "Science tells us that objective nature is nothing but a system of mathematical or rational forms. Hence, there must be a basic Reason at the basis of things. Since this is so, why waste one's time upon the third-rate knowledge of this unimportant physical world of sensation, or even upon the second-rate knowledge of mathematics, which is so difficult to grasp, when by direct communion with this system of rational forms which is in everything and hence in me, I can know reality directly". The result is the Middle Ages. The conception of nature as a system of mathematical forms gives way to the doctrine of a communion of souls; man becomes more interested in securing his eternal salvation in some other world, than in making his self worth saving by means of a disciplined objective study of this one; and science goes into virtual oblivion for centuries.

To be sure, this is the last thing that Plato or the astronomers and mathematicians of his era wanted. But we must remember that it is not what a man thinks personally, but what his official premises imply, that determines the outcome of the acceptance of his philosophy.

If anyone thinks that the relation between the acceptance of the mathematical theory in the Greek period, and the Middle Ages is a purely accidental sequence of events let him consider certain contemporary developments. In his book "Space, Time and Gravitation", the English astronomer and mathematical physicist, A. S. Eddington, went on record for the mathematical theory of nature in the following terms: "The relativity theory of physics reduces everything to relations; that is to say, it is structure, not material, which counts." The concluding chapter of one of his latest books, entitled "The Nature of the Physical World" bears the title "Mysticism". The former treatise was written in 1921, the latter in 1928. It has taken him less than ten years to pass from Plato to the Middle Ages.

Even that great disciple of science, the usually "hard-minded" Bertrand Russell, has been caught by the inevitable logic of this sequence, for the reading of Eddington's book caused him to write an article entitled "The Twilight of Science", in which he heralded the coming of a second Dark Ages. One may well ask, whether science also must have its Judas.

These considerations suggest that the philosophical interpretation which we put upon our science is a matter of no mean importance. First principles seem to have the faculty of producing their logical consequences whether it be in the second century B.C. or the twentieth A.D.

Evidently, it is man's nature, never to give his serious attention to those aspects of his universe which the authoritative scientific philosophy of his age designates as of secondary importance. It is for this reason that Greek science prepared the way for its own decline. One cannot deprecate the importance of the world of sensation and expect science to thrive. Certain contemporary scientists, who are trying so valiantly to convince themselves that their present collection of mathematical formulae, which possess no physical meaning, constitutes an ideal state of affairs, will do well to reflect upon this bit of history.

Thomas Aquinas and The Rebirth of Science

The story of the revival of science in the Western World is most interesting. Sarton has noted[4] that the static character of Eastern civilization is bound up with its failure to throw off the dogma into which its science first crystallized. Many reasons have been given for the different turn which events took in the West. Undoubtedly many secondary causes contributed. Among these were changing political and economic conditions, such as the fall of Constantinople and the coming of the Arabs to Spain. Both of these occurrences brought Europe into contact with the original Greek ideas and texts. It is to be noted that it was the Arabs and not the Christians who preserved Greek science for the European world. Also, there were many scientists, under cover of the darkness, quietly carrying on the pure Greek tradition. But the sum of all these factors does not account for the real miracle. The listing of them does not answer the real question. Why did man regain the interest in nature which would cause him to take advantage of these changes?

Gilson[18] and Taylor[10] have suggested that the answer is to be found in the work of Thomas Aquinas. What, you may ask, could this most abstract of Catholic theologians have had to do with the rebirth of science. The point is that he shifted the theology of the Catholic Church from Platonism to Aristotelianism. A comparison of the epistemological principles of the mathematical and functional theories will reveal the significance of this act. Whereas a belief in the mathematical theory of nature led men to regard observable nature as of only secondary importance, a belief in the functional theory necessitates that one regard the world given in observation as containing the real world. The Aristotelian philosophy does not permit a "bifurcation" of nature. This makes a knowledge of nature for its own sake of first-rate importance. Such a faith, taught in the leading intellectual centers of Christendom, made science inevitable. The change which it wrought would be far more effective than attacks, such as Bacon's, from without, because it had the blessing of the authorities upon it. Thus, it happened that the functional theory of nature of Aristotle became the dominant philosophy of the Western world during the few centuries just previous to the seventeenth.

Galilei and The Dominance of the Physical Theory of Nature

But one cannot bring back an interest in nature for its own sake, and be sure of preserving traditional first principles. The methods of science have no respect for tradition. Evidence soon came to the attention of the Catholic Galilei which could not be reconciled with the functional theory of first principles. This, more than the much-heralded Copernican revolution, is what shook the world. For whereas Copernicus and Kepler introduced conceptions which merely modified the mathematical and functional theories, Galilei's founding of the "Science of Local Motion"[19] necessitated the rejection of those theories. For the second time in the history of Western civilization the first principles of science were in question, and it was evident that a new philosophy must take the leadership of man's thought.

This distinction between scientific discoveries which, while radical for technical science and popular thought, entail merely a certain modification in the development of traditional first principles, and those which necessitate a rejection of the first principles themselves, is very important. It marks the difference between scientific discoveries which are of fundamental importance for philosophy, and those which are not. The helio-centric theory, as it came from the hands of Copernicus and Kepler, and Galilei, the astronomer, brought about a radical shift in technical scientific laws and in man's outlook upon his universe, but it was perfectly compatible with either of the traditional ruling philosophies of science. In fact, both Copernicus and Kepler [20] were ardent adherents of the mathematical theory of nature. It was the ideas and discoveries of Galilei, the physicist, which produced the real revolution—the philosophical revolution, for they necessitated the rejection of the traditional dominant theories of first principles. They were incompatible, not merely with Platonic and Aristotelian science, but also with the Platonic and Aristotelian, or mathematical and functional, philosophies of science. It is for this reason that Galilei ushered in a new day, rather than a mere radical modification in the secondary principles of an old one, when he made forces and the motions of masses the fundamental concepts of science.

His discoveries came to articulate and formal expression in the mechanics of Newton. This new philosophy [21] had two characteristics. It was a physical theory, i.e., it conceived of nature as a system of masses and forces; and it was a mechanical theory. This necessitates that nature must be kinetic atomic in character. For the mechanical aspect of the theory presupposes the validity of the principle of being, and the thesis that the real is being and is physical implies the kinetic atomic theory, as the Greeks proved.

This consequence of Newton's mechanics did not appear explicitly in his statement of its principles. His treatises indicate, however, that he accepted the kinetic atomic theory and hoped that eventually all natural phenomena would be stated in terms of it. However, Liebnitz noted that the principle of the conservation of momentum is unintelligible before the obvious appearance and disappearance of motion unless one accepts the kinetic atomic theory; [22] and later, Laplace found it necessary to state the laws of Newtonian mechanics explicitly in kinetic atomic terms.[23] Thus, with the discoveries of Galilei and the generalization of those discoveries by Newton and Laplace, the physical theory of nature became the dominant philosophy of the Western World, for the first time in its history.

But it was a revised and strengthened physical theory that appeared in the seventeenth century. A weakness in the physical theory of the Greeks had been indicated by the Sophists. They had said, "Even if the atoms exist, what difference does it make?". The point was that the physical theory of nature had little fertility when referred to the concrete instance. Galilei changed this. With his formulation of "The Science of Local Motion" the physical theory of nature was stated in terms of the near at hand. In this, the difference between the Ancient and the Modern world consists.

An examination of Galilei's procedure will make this clear.[24] Instead of centering attention, as did Greek inorganic science, upon the distant regions of astronomical space and the perfect mathematical forms which were supposed to make appearances in those regions intelligible, Galilei took hold of a local ball here in this supposedly insignificant world of sensation, and allowed it to fall to the ground. Desirous of determining the law of its fall, he noted the obvious factors upon which its final velocity might depend. They were, the weight of the ball, the distance, and the time. Thus, three hypotheses were suggested by immediate evidence. The velocity might be proportional to the mass, the distance, or the time. The experiment in which masses of different weights were dropped from the tower of Pisa, eliminated the first hypothesis. He was convinced, without complete justification, that deductive reasoning eliminated the second. Only the third remained. He proceeded to prove this experimentally, by deducing the relation between time and distance which must hold if it is true, and establishing the existence of this relationship by means of his famous experiment with a ball on an inclined plane.

With this verification a new conception of force came into scientific theory. Previously, it had been conceived, in statics, as pressure or displacement, and in Aristotelian dynamics, as that which produces motion or velocity. After Galilei it was to be regarded as that which produces a change of velocity, or acceleration.

This new conception of force provides the key to Galilean and Newtonian mechanics. It had two important consequences.

In the first place, it made the concept of time of great importance. Henceforth, no force could be determined or motion understood except in terms of the change of velocity with time. As Bergson has been quoted as saying, "The concept of time came down an inclined plane from heaven to the modern world, through Galilei". This point is important and somewhat surprising. If time and becoming are such obvious and elemental factors of nature as certain current thinkers have suggested, why did these ideas fail to make their appearance in a significant form, in the philosophy of becoming of Aristotle? Why was it necessary to wait until the founding of modern science for the discovery of the importance of the concept of time? The fact that it did not arise until technical science reared itself upon the physical theory of nature suggests, at least, that time must find its basis in matter and its motion.

Secondly, Galilei's conception of force implies the principle of inertia. This principle affirms that a body, not acted on by external forces, will either remain at rest or move in a straight line with a uniform velocity forever. Galilei gave a somewhat independent intuitive justification for this principle. It is to be noted, however, as Mach has indicated,[24] that it is a necessary consequence of the doctrine that force produces a change of velocity, or acceleration. For if this be the case, it follows, when a force ceases to act, that the body will not stop, but will merely cease to change its velocity. Hence, if it is at rest when the force ceases, it will remain at rest, and if, in motion, it will continue to move with the final velocity which it possessed when the force ceased to apply.

Certain consequences follow from the principle of inertia. In the first place, a definition of straightness must be given, and in the second place, matter is dynamic and dualistic in character.

Since Euclidean geometry was the only one known at the time, it was natural that its metric should be taken as the criterion for straightness. The primacy of the principle of inertia suggested also that the space which it presupposes is absolute. Thus modern physics found itself in accord with Greek pre-Socratic philosophy upon the doctrine that matter moves in space.

The dynamical theory of matter, which the principle of inertia entails, has been overlooked by many critics of traditional modern scientific theory. It has often been maintained, particularly by writers on biological theory, that the dynamical characteristics of nature are inexplicable in physical terms. Matter, they say, is inert. But this charge involves a misrepresentation of the principle of inertia, and a confusion of the Galilean and Newtonian with the Aristotelian or Cartesian theory of matter. It is only in Aristotle's or Descartes' physics, that matter is inert. In the case of Aristotle, this is a necessary consequence of his doctrine of formal and material causes. In order to make this doctrine intelligible, he had regarded matter and form as mere passive attributes of a more dynamic type of substance, the activity of which is the efficient cause or force. Obviously matter which is a passive attribute of a more fundamental substance is inert. It can move only when a force acts, and it must cease to move when the force ceases. Aristotle made this a law of motion in his physics. It was this law which Galilei's new conception of force negated. Matter does not stop moving when a force ceases to act upon it; it merely ceases to change its velocity. Hence if it is at rest, it will remain at rest; if it is in motion, it will persist with a uniform velocity. Obviously, only matter which is sufficiently dynamic to be able to move of its own nature can do this.

When one notes that external forces have their seat in matter, it becomes evident that matter, as conceived by traditional modern science, is not unlike human nature. It tends to maintain uniform motion in itself and accelerated motion in its neighbors. This dualistic and dynamic character was implicit in Galilei's demonstration that force is that which produces acceleration, and in Newton's first two laws of motion.

It remained for the latter to guide these ideas to a tremendous victory. For he took them as demonstrated by Galilei, for the near-at-hand, and passed out through the latter's analysis of the projectile, and Huygens' analysis of centrifugal and centripetal forces, to the moon and to Kepler's three laws of planetary motion, to demonstrate, both experimentally and mathematically, that the physical conceptions, which Galilei had proved to be at the basis of the falling of a ball from one's hand to the ground, are at the foundation of the motion and stability of the entire astronomical universe.

With Galilei the physical theory of nature was stated in terms of the near-at-hand, and with Newton the physical categories of the near-at-hand were revealed to be the key to the far-away. Is it any wonder that modern science is experimental, and that intelligent modern men are absorbed in noting the effects which accrue from the movement of local physical things? Need one ask why the laboratory of the physicist has replaced the Academy of the pure mathematician, or why the this-worldly industrialism of our contemporary life, with its control of the most commonplace practice by the most abstract of experimentally verified scientific theory, has replaced the otherworldliness of the Middle Ages? To all these questions the answer is the same. With Galilei and Newton the mathematical and functional theories of nature were replaced by a physical theory of nature which was made explicit in terms of the near-at-hand.

Nor was this all. After bringing the universe down into the hands of man, Galilei gave him principles connecting effects which touched his happiness, with local masses and forces which were within his control. Thereby, a new conception of fate and freedom took hold of the mind of man. No longer is it necessary, in every instance, to endure natural pestilences which one does not like. Providing that they are connected by the laws of physics to factors which man can control, they can be curbed or removed. This is a new idea. Before Galilei and Newton, all that scientific philosophy did was to enable one to look at events from the "higher standpoint". If anything existed in his immediate environment which man did not like, all that he could do, by the aid of science, was to view it in the light of "the idea of the Good", and learn to grin and bear it. With Christianity the actual result was no different. Although certain sects believed in the freedom of the will, and grand promises were made about the capacity of man to move mountains, there were always certain conditions attached, which no man ever succeeded in fulfilling. The modern philosophy of science changed this. Factors in one's immediate environment were connected to masses which man could take hold of, and were demonstrated to be removable by certain explicit conditions which were well within the capacity of man to fulfil. As Bury has indicated,[25] the idea of progress is a distinctly modern conception.

Note how the revised physical theory of Galilei and Newton has worked itself into the very soul of our present civilization. Its four fundamental concepts, force, mass, space and time exhibit themselves in every phase of our thought and action. It was not an accident that Western mankind stood spell-bound, in worshipful attitude, when the plane of Lindbergh came to rest upon the muddy ground of Le Bourget. For in that act, the ideas which the modern world has regarded as most fundamental, appeared in one of their most spectacular exemplifications. A *mass* had been accelerated by *force* through *space* and *time,* to bring man to one of his greatest triumphs over his environment. One was reminded of that earlier and more difficult flight of the scientific imagination, the forerunner of this one, in which Newton piloted the ideas of Galilei out across the vast regions of astronomical space to demonstrate that the principles conditioning the falling of a local ball to the ground are at the basis of the entire universe. Verily, the physical theory of nature, first discovered by the ancient Greek philosophers, and made explicit by Galilei and Newton in terms of the near-at-hand, is the philosophy of the modern world.

FROM NEWTON TO THE YEAR NINETEEN HUNDRED

The interest in the near-at-hand which Galilei created, gave birth to a second period of specialization in science. Many new movements occurred, four of which must concern us briefly. They involve the sciences of chemistry, biology, thermo-dynamics, and electro-magnetics.

Modern chemistry began with Boyle but came upon discoveries which necessitated a new theory of first principles, because of the work of Priestley, and, particularly, Lavoisier. The latter demonstrated that heating will transform the silver-colored, semi-liquid substance, mercury, into a powder of an entirely different colour, which, if heated with carbon, can be turned back into the silvery, semi-liquid mercury again. Here was an obvious change of substance as well as properties which left one at the end of the experiment with the original materials. Furthermore, by taking advantage of Priestley's discovery of oxygen, and by following gases with the use of the balance and sealed containers, he was able to demonstrate that the weight remained constant through these marked changes of property and substance. Thus the principle of the conservation of mass was discovered.

But the significance of this experiment was not exhausted at this point. For it left the science of chemistry face to face with the same two facts of stuff and change which the Greek pro-Socratic or inductive natural philosophers had found to be extensive characteristics of nature. Mass or stuff was conserved (the principle of being) through chemical change. It is a testimony to the soundness of Greek thought and the physical theory of nature, that the result in both cases was the same; for soon after Lavoisier, Dalton found it necessary to rear the science of chemistry upon the atomic theory. Thus independent evidence in chemistry combined with that of Newtonian mechanics to reconfirm the conclusions of Greek inductive natural philosophy concerning the validity of the physical theory of nature.

The development of the science of chemistry beyond Lavoisier and Dalton is well known. Time does not permit a summary of its many accomplishments here. However, one new factor does appear which must be mentioned. In the nineteenth century Mendeleev discovered that a certain structure of matter exhibits itself in the properties of the chemical atoms. The periodic table of the elements is an expression of this fact. It is important not merely because it provides the basis for verification of many of our current theories of atomic structure, but also because it is an exhibition within chemistry of the general fact of order and relatedness in nature which will be found to be making itself more and more evident in every branch of scientific endeavor.

In biology a similar development took place. Before the seventeenth century the teleological concepts of Aristotle and Galen dominated this field. The first break from this tradition came when Vesalius took his life in hands to reject traditional ideas, and found anatomy upon the description and dissection of actual bodies, rather than upon the authority of the text-books. Soon after this Galilei was lecturing at the University of Padua to audiences of two thousand on the foundation of mechanics. A comparison of dates will indicate that the English physician, Harvey, was there studying medicine when these lectures were being delivered. He must have been influenced by them. This circumstance becomes important the moment one notes that the discovery of the circulation of the blood, which Harvey announced after his return to England, involves the first application of mechanical principles to gross physiology. The application of physical and mechanical conceptions to living as well as inorganic systems was, thereby, demonstrated.

This is but the beginning. Before the end of the eighteenth century Lavoisier began to think about living things from the chemical and the thermodynamical point of view. By experimental means, in conjunction with Laplace, he was able to establish a quantitative identity between the heat derived from the metabolic transformation or burning of carbo-hydrates in the body, and the burning of an equivalent amount of carbon outside the body. Thus, in the mind of this great Frenchman, not merely chemistry, but the sciences of physiological chemistry and bio-physics were born, and the application of physical and atomic categories to rabbits and men, as well as planets and pendulums, was demonstrated.

The genius of this martyr of the French Revolution has not been appreciated. For a well balanced conception of living things in their relation to inorganic nature there are few if any to compare with him. He noted the reciprocal relation which exists between plants and animals, —the plants synthesizing carbohydrates and giving out to animals the oxygen which the latter use to burn up these carbohydrates taken in as food; and the animals breathing out carbon dioxide which the plants absorb. In other words, Lavoisier put his finger upon the oxygen-carbon dioxide cycle.

In addition he noted, although the idea of energy had not been explicitly formulated, that the dynamic attributes of living creatures involve, in some fundamental sense, exchanges of energy, which make them dependent upon, and an integral part of, the inorganic world. He had the idea that energy comes into living systems from the outside, and there are suggestions that he was aware of their dependence upon energy from the sun. These are important considerations to which we must refer in a later chapter. The point to note now is that after Lavoisier, it was very difficult for any informed thinker to escape the conclusion that a living thing is a physico-chemical system.

This conception received a rather startling confirmation when Wöhler produced certain products of living things in a laboratory, and Liebig demonstrated the application of the physical and atomic categories to plants, by placing the science of agricultural chemistry upon a secure and well-organized basis.

Meantime descriptive biology had picked up the materials and principles of biological classification where Aristotle and his successors had dropped them. It grew with the work of Linnaeus, De Jussieu, and Cuvier, until Darwin began to note the relationship between living things in the light of their distribution in space and time, and was led to the theory of organic evolution and the rediscovery and verification of the mechanical principles of struggle for existence and survival of the fittest, which Empedocles had introduced long before. With the work of Darwin the Aristotelian doctrine of fixed and irreducible forms received a conclusive rebuttal from the very science in which it arose; types are not fixed, forms cannot be regarded as ultimate, and irreducible to physical and mechanical conditions. What had been made evident long before to informed minds by Galilei, became inescapably apparent to all with Darwin. Man as well as motion must be conceived in the light of the acceleration and change which time introduces. Even his origin is subject to temporal and ruthlessly physical and mechanical conditions. With Darwin the physical theory of nature is stated in terms of the near at hand to a degree which makes it almost impolite, and unduly personal.

In this fashion the physical theory made its way in the

organic as well as the inorganic world. In this new field also, the fact of relatedness appeared, as it did in chemistry with Mendeleev. The modern scientist who emphasized this for biology was the French physiologist, Claude Bernard. Perceiving that the living organism is a physico-chemical system, and that it must be approached from such a point of view, he insisted that physiology must be placed upon an experimental basis.[27] At the same time he pointed out that the physico-chemical materials of the body enter into an organic relationship which is the distinguishing mark of living things. To emphasize the former to the neglect of the latter is to fail to touch life itself. Thus biology was brought face to face again with the two facts of mechanical causation and organization which Hippocrates of Cos had noted at the time of its origin in ancient Greece; the problem of organization remained unsolved. It appeared that the old scientific and metaphysical problem of matter and form had been discarded from descriptive biology by Darwin, only to turn up again in physiological chemistry with Claude Bernard. Nevertheless, the evidence of modern biology combined with that of chemistry and mechanics to reconfirm the validity of the physical theory of first principles. This was certain. Yet something else also seemed to be true. There were signs of an organization and relatedness in living things which traditional physical categories seemed incapable of conditioning. This was the first sign that something was lacking in the theory of first principles of modern science. The warning was ignored, however, because the evidence for the validity of the physical theory of nature was overwhelming, and there were no signs of any weakness in this theory in physics.

The third major movement to which we must refer occurred in the sciences of optics and electricity. Huygens and Newton had given evidence in support of both the emission and wave theories of light. The latter was fully aware that neither theory accounted for all the known facts. Unfortunately, however, the authority of his name became associated with the emission theory, with the result that the wave theory won its way only with great difficulty. The accepted ideas of mathematical physics also contributed to this end. The mathematicians treated nature as discontinuous. The evidence for atomicity and discontinuity was overwhelming. It seemed logical, therefore, that light must involve an emission of particles. However, at the opening of the nineteenth century an Englishman named Young made a study of the phenomenon known as interference, introduced the wave theory to account for it and previously unexplained phenomena, and postulated a luminiferous ether, "rare and elastic in high degree" which pervades the entire universe.[26]

But the wave theory won little acceptance until Fresnel of Paris took up Young's ideas and extended them to account for polarization. He met immediate opposition from the authoritative scientists of the French Academy. They decided to put an end to all these arguments for the wave theory, by offering a prize for the best paper on the subject. Fresnel entered the competition, reviewed both theories more thoroughly than had been the case before, and demonstrated that the wave theory has the greater justification. But the authorities were not convinced. In fact, Poisson who was one of the leading mathematical physicists of the time, and a judge in the contest, charged that Fresnel's theory was false since its truth entailed certain consequences which were absurd. It occurred to someone to check these deductions of Poisson; it was found that they exist in fact. Thus the opponents of Young and Fresnel were condemned by their own arguments, and the wave theory of light, with its attendant luminiferous ether, became an accepted part of modern scientific theory.

Soon after this, experimental study of electricity and magnetism led to a somewhat similar conclusion. As a result of observations of the distribution of iron filings around the poles of a magnet, and other electrical and magnetic phenomena, the great English experimental physicist, Faraday, was forced to the conclusion that electro-magnetic phenomena depend upon the field in which the bodies in question are imbedded as well as upon the properties of the bodies themselves. His reflections expressed themselves in the theory that there are lines of force extending out through all space. Again, the fact of continuity was forcing itself upon science. Since previous laws in electro-magnetics were stated in terms of a discontinuous theory of nature, Faraday's ideas received no attention from the authoritative mathematical physicists of the time. It remained for a young and brilliant mathematical student, named Clerk Maxwell, to come down from Edinburgh to Trinity College, Cambridge, where he took up Faraday's purely physical conceptions and gave them the exact mathematical formulation which they needed. He soon demonstrated that all the traditional laws could be stated in terms of the continuous field theory which Faraday had proposed. Furthermore, it became evident from his mathematical analysis that the field or luminiferous ether which the wave theory of light demanded was identical with the field or electro-magnetic ether which Faraday's lines of forces required. Thus optics was revealed to be but a part of electro-magnetic theory. This conclusion was confirmed experimentally by Hertz, a teacher in the Technische Hochschule of Carlsruhe, Germany. Later, it was taken up by Marconi and Edison, to become the foundation of the electrical age in which we now find ourselves. It cannot be too strongly emphasized that the day of leadership of the purely practical man has passed. For the commonplaces of practical life rest upon principles which are so unobservable and unexpected that only the most refined mathematical computations and the most abstract of theories can express them. Since the time of Maxwell nothing has been more practical than the scientifically theoretical.

But the electro-magnetic theory of Faraday and Maxwell has even greater philosophical significance. It led to the introduction into scientific theory of a continuous physical substance, the ether. The question immediately arose concerning the relation between this ether and the discontinuous atoms of chemistry and traditional physical theory. The attempt was made by Kelvin and others to reduce the discontinuous to the continuous by regarding the atom as a mere vortex or pucker in the ether. This attempt was successful for electro-statics, but broke down, as anyone acquainted with the analysis of Parmenides would expect, when applied to electro-dynamics. For when Lorentz generalized Maxwell's equation so that they applied to electrified or magnetized bodies in motion he was forced to an atomic theory of electricity. Later Sir J. J. Thomson isolated the electron experimentally and Millikan determined its charge. Thus, notwithstanding the introduction of the continuous ether, electro-magnetics joined with biology, chemistry, and Newtonian mechanics to reconfirm the validity of the discontinuous physical theory.

The final modern scientific movement, to concern us, produced the sciences of thermo-dynamics and statistical mechanics. It had its origin in many different sources. Boyle's law relating the volume and pressure of gases, Count Rumford's discovery and Maxwell's and Clausius' development of the kinetic theory of heat and gases, and all the experimental work done by men like Faraday, Volta, Lavoisier, Mayer and Joule which indicated an equivalence between thermal, mechanical, chemical, magnetic, and electrical energy, entered into the final conception. From all these findings the idea gradually arose that there is something having the capacity to do work which is other than matter, and which remains the same during all transformations. The word energy was finally coined to denote this, and the principle, that this thing termed energy is neither created nor destroyed by any transformation of it from one form to another, was finally discovered and announced by Mayer in 1842. This principle of the conservation of energy is known as the first law of thermo-dynamics. It remained for Joule to put it on a sound quantitative basis by determining the mechanical equivalent of heat, and for the great German physicist Helmoltz to sense its full theoretical importance.

Eighteen years earlier, in 1824, a Frenchman, Sadi Carnot, discovered, in connection with studies of the efficiency of the heat engine, that in an isolated physical system, the amount of energy which is available for work is constantly decreasing. The generalization of this discovery is known as the second law of thermo-dynamics. Clausius defined it as the principle that the entropy of the universe, or the amount of energy unavailable for work, tends toward a maximum. When the first law of thermo-dynamics was discovered it was thought that these two principles conflict. But Helmholtz soon showed that the validity of the second law is necessary in order to reconcile the first law with the impossibility of perpetual motion. Were it not for the second law we could have a machine which would run automatically without fuel, and thus be able to get work done for nothing.

Two points concerning the science of thermo-dynamics are significant. In the first place, its first law did not become plausible until heat was conceived in kinetic atomic terms. If heat is nothing but the motion of microscopic particles then the equivalence of thermal energy with the mechanical motion of molar objects becomes intelligible. Thus it may be said that this science gains its plausibility from, and hence, in a certain sense reconfirms the validity of, the kinetic atomic theory. Secondly, it has been usual since the time of Boltzmann and Gibbs to state the science of thermo-dynamics in statistical terms. From this point of view the second law means that the universe is proceeding by the laws of chance, as defined by the theory of probability, from an original improbable state of complicated organization and non-uniformity of temperature to a more and more probable state of uniformity and homogeneity of structure and temperature. This is one of the most unethical and discouraging laws, from the point of view of human interests and values, in the whole of science, for it means that the universe in which we live is headed to a complete destruction of us and our solar system with an utter indifference to human values. It is also one of the most fascinatingly important and elusive laws in science. There can be no doubt of its truth in some fundamental sense. The specific principles, which it provides man in his study of chemical changes, have been confirmed again and again. The monumental work of Willard Gibbs, in the founding of physical chemistry, rests upon it. It is true in some sense, but there are difficulties also. We must return to them in a later chapter.

Suffice it to say now, that with the statistical interpretation, the sciences of thermo-dynamics and statistical mechanics merge and the discontinuous theory of nature receives an additional confirmation. This atomic or discontinuous aspect of thermo-dynamical and statistical theory comes out most explicitly in the contributions made at the beginning of this century by the German physicist Planck,—contributions which form the basis of current quantum theory and wave mechanics.

A failure of thermo-dynamical theory to correspond to fact in certain of its deductive consequences led to a reexamination of the foundations of statistical theory. These studies led not merely to the rearing of statistical theory upon an unequivocally discontinuous basis, but gave rise to a certain constant, known as Planck's constant, which led to the conclusion, when introduced into thermodynamical theory, that energy, as well as matter and electricity, is atomic in character. In this manner, the sciences of thermo-dynamics and statistical mechanics reconfirmed the validity of the discontinuous physical theory of nature.

Such was the state of scientific theory at the opening of our century. The physical theory of nature remained in the same inescapable connection with the obvious extensive facts of stuff and change which the Greeks had indicated, and in addition it had been reconfirmed by independent evidence in each of five different branches of modern technical science: mechanics, chemistry, biology, electro-magnetism, and thermo-dynamics. How anyone can review this tremendous accumulation of evidence and verified theory dating from the seventh century B.C. to the present moment, and not accept the kinetic atomic theory is difficult to understand.

Nevertheless, a careful consideration of the state of scientific theory at the opening of this century should have raised certain questions. First, the writings of Claude Bernard on physiology had revealed that the problem of biological organization had not been solved. There appeared to be a form in living things which the traditional physical theory seemed inadequate to fully explain. Secondly, the physical atomic theory had been able to preserve itself before the increase of exact information, only by complicating its doctrine by the addition of absolutes. In the Greek period the fact of motion led to the introduction of absolute space. With Galilei and Newton absolute time and absolute gravitation appeared. And with Maxwell the absolute ether was admitted in good standing. The presence of these absolutes should have made men suspicious, for they represent the two facts of continuity and relatedness before which the kinetic atomic theory is weak. This theory maintained that nature is discontinuous and that all relations are variable effects of atomic motion. Yet nature revealed uniform temporal and spatial and gravitational relatedness, and the need of a field, possessing continuous attributes, for the transmission of electro-magnetic waves. Facing these facts man kept his kinetic atomic theory as it was, and added on new absolutes to take care of each of the difficulties which appeared. But was this procedure correct? May it not be that the kinetic atomic theory itself needs modification so that it can account for continuity and relatedness? May not the addition of these absolutes constitute the preservation of an original error by the patching up of the traditional theory, when what is really called for is a modification in the kinetic atomic theory itself?

The analysis of space-time, quantum phenomena, and biological organization which will appear in the next three chapters indicates that this is the case.

EINSTEIN AND THE FIRST PRINCIPLES OF SCIENCE

Meantime, however, Einstein, motivated by slightly different considerations which arose in electro-dynamics, appeared with his revolutionary conceptions. With one stroke, in the discovery of the special theory of relativity, he removed absolute space, absolute time, and the absolute ether to substitute another absolute called space-time; and with another master stroke in the general theory of relativity, he removed absolute space-time and absolute gravitation to leave science with nothing but matter and a certain mathematical formula termed the tensor equation for gravitation, which specifies how the properties and distribution of matter determine the metrical structure of a variable and hence relative four-dimensional Riemannian space-time.

Consider, the effect of the special theory. We have noted that scientific evidence makes the acceptance of the physical theory of nature inescapable. We have noted also that this theory is untenable unless there is a referent for atomicity and motion in something other than the microscopic particles, and that this referent was identified with absolute space. The first significant effect of the special theory of relativity is the removal of absolute space. It becomes evident, therefore, that the physical theory of nature which provides the basis of modern science and modern civilization has been shaken to its very foundations. It is left without any meaning for atomicity and motion. Certainly a physical theory which is unable to account for the obvious fact of change, and which cannot provide science with such necessary notions as atom, motion, or momentum is of no use. Of one thing we can be certain at the very outset, because of our knowledge of the alternatives which the Greeks indicated:

either the traditional physical kinetic atomic theory must be radically amended to provide a new referent for atomicity or motion, or the science and civilization of our own day must be reared upon entirely different philosophical foundations from those of the last three centuries. In either event the consequences are most significant, not only for science and philosophy, but for each and every branch of human interest and activity. This is the general significance of the theory of relativity. It has brought the philosophical foundations of modern science and the modern world into question.[28]

This exhibits itself in another problem to which Einstein's theory has given rise. It appears in connection with the general theory of relativity. The essential thing in this theory is the tensor equation for gravitation. Its extremely mathematical emphasis has led Eddington to conclude that physics reveals only mathematical relations to be ultimate. But this equation necessitates that the metrical structure of space is conditioned by the motion and distribution of matter. Hence, we can understand why Einstein maintains that space-time must be defined in terms of matter. However, Professor A. N. Whitehead has pointed out that it is very difficult to understand how measurement is possible if this is the case. The justification for this contention must wait until the next chapter. Suffice it to say that it leads him to the conclusion that a space-time structure or form exists in nature which is independent of matter.

Consider this situation in the light of the history of Western science and civilization. In the Greek period three different theories of the first principles of science arose out of its empirical investigations, to reveal the fundamental metaphysical problem of matter and form which is their common basis. In the succeeding centuries up to the present moment, each of these three theories has had its chance to determine what man regards as of first importance, and thereby to dictate his conduct. The mathematical theory in a degenerate and Christianized form came first, the functional theory second, and the physical theory third and last. These periods of dominance correspond to the Dark Ages, the Scholastic Period, and the Modern Era, respectively. A comparison of the main characteristics of each age with its corresponding theory will reveal that the connection is not purely accidental. The mathematical theory can not make sense unless one deprecates the importance of the world of sensation. The Dark Ages with its neglect of empirical science and its other-worldly interests is such a philosophy in practice. The functional theory insists that the real is contained in sensation. The Scholastic period is one in which an interest in nature for its own sake revives. The physical theory of modern science taught that masses and forces, understood and controlled by means of laws discovered by an experimental study of the near at hand, are the key to everything else. The modern era of technical science and theoretically controlled practice is the logical outcome of such a belief. Science and common sense have supposed that the answer of the modern world is decisive and final. But what do we find now? The physical theory has been shaken to its very foundations. It stands without any theory of atomicity or motion and without an adequate theory of measuring. Furthermore, in raising within physics itself the vexing problem of the relation between matter and space-time which divides Einstein, Eddington, and Whitehead, the theory of relativity has brought us back again to the same old problem of matter and form which the Greeks left. For what is space-time but a system of mathematical relations, and what is mathematical relatedness, but the old form of Greek science and philosophy.

This throws an entirely new light upon the nature of the course of Western civilization, and upon the degree of finality of modern thought. In fundamental matters we have not gone beyond the Greeks. For we are still facing their problem. Moreover, the three different answers which contemporary scientists have given to it are the same three theories of nature which the Greeks proposed. Einstein, in his original papers, would define space-time in terms of matter. This is the physical theory of nature. Eddington would reverse the relationship. This is the mathematical theory. And Whitehead would regard matter and space-time as abstractions from a monistic process of becoming. The identity of this with the functional theory is evident. Apparently science has become so new that it is very old.

Furthermore, Einstein's discovery that it is impossible to define the motion of matter in terms of a relation to space is not particularly original. It was pointed out in ancient times by Zeno. Notwithstanding all the evolution and revolution which moderns have emphasized, we have been simply adding our particular bit of new information and trying out our important inadequate answer to the same old problem. Moreover, our theory, like theirs, must be transcended. The final result should be a philosophy which combines the intensive technique and ill-proportioned knowledge of the Moderns, and the logical rigor of the Scholastics, with the extensive knowledge, the clear and deep thinking, and the remarkable sense of proportions of the Greeks. It is because contemporary scientific discoveries have prepared the way for such an advance, by bringing the first principles of science into question, for but the third time in the history of the Western world, that they are of such exceptional significance.[28]

But if the problem which we face and the solutions which are being offered are the same, at bottom, as those which the Greeks discovered, the information available now, as clues to an adequate answer, far transcends theirs. This, together with the important knowledge that traditional theories of first principles are incomplete or inadequate, is the real and permanent advance which we have made over the past.

Furthermore, history indicates that sooner or later a new theory of first principles will crystallize out of the present flux of facts and rival interpretations, and that when it comes it will determine the character of civilization in the centuries immediately ahead. The philosophy of science does not change very often, but when it is altered, everything from the foundations of human activity to the peak of religious faith is transformed. Also, unless we pass to a more fundamental issue, and no such transition seems to be in evidence or called for, the new universe which we shall construct within and around ourselves must rest upon some form of one or another of the three bases which the Greeks outlined. Anything else can mean only muddle-headedness and confusion.

The most important question now arises. What theory of the first principles of science will gain this ascendency? All three theories have proved themselves to be inadequate. It appears that a revision in one of them is necessary. What must be the nature of this revision? An answer to this question necessitates an understanding of current scientific conceptions. We shall enter upon such an understanding in the next chapter, with an exposition and analysis of the specific principles and consequences of the theory of relativity.

References and Bibliography

1. C. M. Bakewell. Source Book in Ancient Philosophy. Scribners.
2. J. Burnet. Greek Philosophy: Thales to Plato. Macmillan.
3. L. Robin. Greek Thought. Knopf.
3[1]. Lucretius De Rerum Natura.
4. G. Sarton. Introduction to the History of Science. Vol. I. Williams and Wilkins.
5. The Legacy of Greece. Edited by R. W. Livingstone. Oxford Press.
6. T. L. Heath. The Thirteen Books of Euclid's Elements. Cambridge Press.
7. T. L. Heath. Aristarchus of Samos. Oxford Press.
8. The Dialogues of Plato. Trans. by Jowett. Oxford Press.
9. A. E. Taylor. Plato, The Man and His Work. Methuen.
10. A. E. Taylor. Platonism. Marshall Jones.
11. The Genuine Works of Hippocrates. Trans. by F. Adams. London.
12. W. E. Leonard. The Fragments of Empedocles. Open Court.
13. The Works of Aristotle. Trans. into English. Ed. by Ross. Oxford Press.
14. W. D. Ross. Aristotle. Methuen.
15. A. N. Whitehead. The Concept of Nature. Ch. IV. Cambridge Press.
16. F. S. C. Northrop. Theory of Relativity and First Principles. Jr. Phil. XXV.

17. C. Kingsley. Hypatia.
18. E. Gilson. La Philosophie au Moyen Age. Payot.
19. G. Galilei. Two New Sciences. Trans. by Crew & De Salvio. Macmillan.
20. E. A. Burtt. The Metaphysical Foundations of Physics. Harcourt Brace.
21. I. Newton. The Mathematical Principles of Natural Philosophy.
22. Leibniz. The Monadology. Trans. by Latta. P. 93. Oxford Press.
23. P. Laplace. Mécanique Céleste.
24. E. Mach. The Science of Mechanics. Open Court.
25. J. B. Bury. The Idea of Progress. Macmillan.
26. J. Merz. History of European Thought in Nineteenth Century. Blackwood.
27. C. Bernard. Experimental Medicine. Macmillan.
28. F. S. C. Northrop. Philosophical Consequences of Theory of Relativity. J. Phil. XXVII.
29. E. Meyerson. Identité et Réalité. Alcan.
30. J. Perrin. Atoms. Constable.
31. A. N. Whitehead. Science and the Modern World. Chs. I–VII. Macmillan.
32. A. S. Eddington. Space, Time and Gravitation. Cambridge Press.

CHAPTER II

The Theory of Relativity

The theory of relativity is interesting on its own account, wholly apart from its philosophical implications. For sheer beauty of logical structure and dialectical development from known fact to final verified consequence there are few if any theories in the whole of science to compare with it. Hence, the necessity of having to examine it in order to determine the first principles of contemporary science, is a most pleasant one.

We have suggested that the theory of relativity is a continuous development from traditional evidence to new and revolutionary consequences. This development falls into five parts: (1) mechanical and electro-magnetic theory before Einstein, (2) the special theory of relativity, (3) the general theory of relativity, (4) the theories of the finite universe and the unitary field theories, and (5) the necessary consequences of the special and general theories. These divisions must be kept clearly in mind. Experimental and logical considerations enable us to regard the first three stages as certainties. The conclusions reached in connection with the last division, if properly treated, should possess equal certainty, since they merely designate what else must be true if the special and general theories are true. The theories of the finite universe of Einstein and De Sitter, and the unitary field theories of Weyl, Eddington, and Einstein rest on more questionable assumptions and lack the experimental verifications which Einstein's earlier discoveries enjoy. But that theories of this character must exist, there can be no doubt.

The reader will appreciate that in no part of technical science can one be as sure of resting one's philosophy on solid foundations as in mathematical physics. But even in this field, the nearer one can stay to the special and general theory and its inevitable consequences, the more sure of one's ground one can be.

MECHANICAL AND ELECTRO-MAGNETIC THEORY BEFORE EINSTEIN

We noted in the previous chapter that the essential contribution of the mechanics of Galilei and Newton was the statement of the physical theory of nature in terms of the near-at-hand. This made experimentation important, and measurement and quantitative treatment possible. The specific discovery which produced this new procedure was Galilei's designation of force as that which produces acceleration, or a change of velocity. This made time an important concept and gave rise deductively to the principle of inertia, which in turn provided a meaning for mass as that which preserves constant velocity. The principle of inertia also presupposed the existence of absolute Euclidean space in its definition of rest and rectilinear motion. Thus in Galilei's new conception of force four absolutes, mass, force, space, and time, were present. We noted also that matter was dualistic in character. It tends to produce rest or uniform motion in itself and accelerated motion in its neighbors. Science expressed this fact by saying that a body has two masses, the one inertial, the other gravitational. However, Galilei's experiment at the tower of Pisa had indicated that these two masses are quantitatively identical. More precise experiments have confirmed this conclusion. This fact is referred to as the equivalence of inertial and gravitational mass. It is important because it forms one basis for, and receives its first theoretical explanation, in the general theory of relativity. In Newtonian mechanics it was a happy coincidence.

The Newtonian conception of space was somewhat complicated. The presupposition of it in the definition of inertia suggested that it is absolute. Furthermore, the atomism and motion of matter required a referent other than the moving masses. This referent had been identified with space. Thus the notion of space seemed to be implicit in the notion of matter. Newton says precisely this. At the beginning of the Principia he writes: "All things are placed . . . in Time as to order of succession, and in Space as to order of situation. It is from their essence or nature that they are Places." In other words, the idea of a mass apart from some common referent other than that mass is unthinkable. He, like the Greeks, took it for granted that this referent is space. But the Newtonian theory of space does not end at this point. The new conception of force made the determination of velocities essential. But velocities cannot be determined without measuring distances and time, and measuring is not possible without introducing a certain amount of relativity. In order to measure a distance to which one cannot apply a measuring rod directly at least two things are essential. In addition to a physical standard measuring rod, one must have a system of coördinates and this system of coördinates must have its zero point fixed to some molar object. Hence all the measurements of distance which we make are relative; they refer to a particular reference body which is chosen.

The system of coördinates which one uses is determined by the nature of the space in which one's measurements are made. Since Galilei and Newton took it for granted that space is Euclidean, systems of coördinates for Euclidean space were used. Such systems are constituted of three straight lines intersecting at a zero point at which they are perpendicular to each other. Such a system of lines is called a Cartesian system of coördinates. Its characteristic is that any point in space can be uniquely designated by three numbers which indicate the distance of the point in question from the zero point along each of the three axes of the system. A Cartesian system of coordinates may be defined as a three-dimensional system of lines referred to a common zero point, and obeying the rules of Euclidean geometry.

Measuring in Newtonian mechanics also requires that the zero point of such a system of coördinates be attached to some molar body. But the principle of inertia dictated that only certain bodies are permissible. Careful observation indicates that the fixed stars seem to be practically free from the action of external forces. Hence they must appear to any system of coördinates which obeys the principle of inertia as being at rest or moving with a uniform velocity in a straight line. This eliminates the earth, for example, as a suitable frame of reference, for relatively to it they cruise around in a great circle. Thus only the fixed stars, or bodies at rest or in uniform rectilinear motion relatively to them, are suitable frames of reference for Newtonian mechanics. Such frames are known as inertial or Galilean frames of reference. It is essential for an understanding of the difference between the special and general theories of relativity that one knows what is meant by such a frame. It may be defined as a system of Cartesian coördinates attached to a molar body which satisfies the principle of inertia. Frames undergoing rotational or non-uniform motion are known as non-Galilean frames.

It is to be noted that the laws of Newtonian mechanics are valid only for measurements made from or referred to Galilean frames of reference. In other words our classical laws of nature held only if we approached nature from a certain physical standpoint. Such was the relativity which measurement introduced into modern science.

But Galilei observed at the very outset that any inertial or Galilean frame of reference would do. Note what this means. To take a physical object as a reference body is to regard it as at rest. Hence, to assert that one Galilean frame of reference is as good as any other is to maintain that there is no means of distinguishing between rest and motion in the case of bodies that are in uniform rectilinear motion relatively to each other. Consider two bodies *a* and *b* such that one is moving in a straight line with a uniform velocity of forty miles per hour relatively to the other. What Galilei noted is that one can regard either of these bodies as at rest without altering the laws of motion which describe their behaviour. *a* may be taken as at rest, in which case *b* will appear to move relatively to it in a certain direction at the rate of forty miles per hour; or *b* may be taken at rest in which case *a* will appear to move in the opposite direction with the same velocity. This equivalence of inertial or Galilean frames is known as the principle of relativity for Galilean frames of reference. It means that the laws describing a given phenomenon have precisely the same form, whether observations are made from one Galilean frame of reference or another.

It is essential that one grasps the meaning of this principle before proceeding further, for it is one corner-stone of the special theory of relativity. Furthermore, it is to be emphasized that we are now talking about classical mechanics. The principle of relativity is not the theory of relativity, but a commonplace rule of Newtonian science. It was discovered by Galilei, and appeared in Newton's Principles as a corollary to his three laws of motion.

The significance of this principle should be noted immediately. We have indicated how measuring brings in reference frames and introduces relativity into science. Were this the whole truth there could be no laws of mechanics which applied to objective nature. We should have laws for nature measured from one fixed star, and different laws for nature measured from another fixed star, and as many different sciences of mechanics as there are possible bodies in nature which we might use as reference frames for our measurements. It is the principle of relativity which enables us to escape such relativity. It is this principle which enables our laws to transcend the necessary approach to measured nature through coördinate systems and reference frames. It does this by expressing the fact that the laws of nature preserve a constant form through all the different space and time values which measurements from different reference frames reveal. Thus paradoxically as it sounds, it is nevertheless true, that the principle of relativity takes science to the absolute. It does this, not by denying the existence of the relative, but by admitting it and finding something in the form of the laws of mechanics which is common to all the variations which different frames introduce.

It is to be noted, however, that in classical mechanics this principle held only for Galilean frames of reference. We can see now that this was a very undesirable and unsound state of affairs. For it means that the laws of Newtonian mechanics apply, not to objective nature, itself, but to nature referred to certain physical objects called Galilean frames of reference. What, we may well ask, does nature have to do with the particular type of reference body a scientist picks out when he makes measurements? Certainly there must be something lacking in a science which forces one to talk about nature as seen from a certain physical standpoint, rather than about nature itself. It is the great achievement of Einstein to have removed this last remnant of relativity by extending the principle of relativity so that it applies to any frame of reference whatever. Thereby science obtains laws of a form which hold regardless of the physical object we use as a zero point in our coördinate system and as a referent for our measurements. This makes it clear that there is a very real and literal sense in which the theory of relativity is inspired by a quest for the absolute. We have also found one of the respects in which relativity physics is a logical development of classical physics.

But a principle is worth nothing to science unless it is true. The enunciation must bespeak a fact. It is well that we think of this rather abstract principle in terms of the evidence which causes informed minds to believe it. It goes back, like all sound science, to observable factors within the experience of each one of us. Galilei first put his finger upon them. He wrote: "The ship may move with any velocity you please, but if only its motion is uniform and it does not rock from side to side, you will notice no change in any of the things we have mentioned nor be able to judge from any of them whether the ship is moving or at rest. If your friend is standing in the prow and you on the poop, greater force need not be used in throwing something for the other to catch, than if your positions were reversed. Fishes in a bowl do not swim with more difficulty towards the front than the backward part of the bowl, but come with equal ease to look for their food, put on any part . . . of the bowl. Lastly, the butterflies and flies flutter about making no difference (as to direction) and it does not happen that when they are tired with staying in the air they are more likely to settle on the wall looking toward the poop as of following the course of the vessel more swiftly. In the same way we shall see the smoke from a grain of lighted incense rising like a cloud and not moving more to one side than the other."[1] It is evident that the principle of relativity for Galilean frames of reference is a generalization from facts which are within the ordinary experience of every human being. Moreover, no possible experiments in mechanics which scientists have devised have altered man's certainty concerning the validity of this generalization.

This aspect of the principle of relativity for Galilean frames of reference permits us to define it in another way. We have previously stated it as the thesis that the laws of mechanics for a given natural phenomenon possess the same form whether that phenomenon is measured from one Galilean frame of reference or another. To this we can now add the more concrete statement that no possible experiment performed within a given Galilean frame of reference will enable one to tell whether it is at rest or in motion with a uniform velocity in a straight line. This, it is to be noted, is but a summary of what Galilei wrote.

It has an unexpected consequence with reference to the Newtonian theory of space. This brings us back to the concept which initiated our discussion of measuring and frames of reference. We noted that motion and the principle of inertia called for a referent which Galilei and Newton identified with space. To say, however, that there is no means of distinguishing between rest and motion for Galilean frames of reference is to say that there is no single absolute space the same for all the different Galilean frames. If the body *a* may be regarded as at rest then the space of nature is at rest relative to *a*, but if *b*, which is moving uniformly relatively to *a*, may also be regarded as at rest, then the space which is at rest relatively to *b* is moving relatively to *a* and hence is not the same as the space of *a*. Hence the space which one regards as the basis for rest is relative to the particular Galilean frame which one chooses. We find, therefore, that the conception of space in classical mechanics was rather complicated. In general theory space was absolute but in actual practice it was relative for Galilean frames of reference. To this we must add the additional complication that it was regarded as absolute both in theory and practice for non-Galilean or accelerated frames of reference. Newton's experiment with the rotating bucket and the behavior of the Foucault pendulum was interpreted as an expression of the fact that it is possible to determine by an experiment entirely within a body itself whether it is at rest or in accelerated motion. The experience of a jerk received when seated in a train, which enables us to know that the train is moving, without looking out of the window to consider it in relation to some other object, is a case in point.

Such was the theory of fundamental principles and concepts in classical mechanics at the opening of our century. Nature was conceived as a system of masses and forces operating in absolute space and time. Forces were conceived as that which produces a change of velocity. Time was regarded as an endless one-dimensional series of instants, such that if two events were assigned correctly by an one to one member of the series, they must be assigned to the same instant by all other observers who measure correctly. Mass, when considered as acted upon by external forces, was conceived as of two kinds, one the inertial mass which opposes a change of velocity, the other the gravitational mass which is responsible for weight. In theory these two masses of a given body were different; in practice they were quantitatively identical; hence, the principle of the quantitative equivalence of inertial and gravitational mass. Space on the other hand was absolute in theory for Galilean frames and in theory and practice for non-Galilean frames of reference, and relative in practice for Galilean frames. The latter fact, and the partial success in escaping from the relativity which measuring introduces, expressed themselves in the principle of relativity for Galilean frames of reference.

This principle had a different standing in the science of electro-magnetics. Since it was this particular branch of modern science which gave rise to Einstein's discovery, we must pick up the historical threads of it where we dropped them in the previous chapter and follow them through to the year 1905. We noted how the ideas of Young, Fresnel, and Faraday came to a unified mathematical formulation with Clerk Maxwell, and to experimental verification with Hertz. In these two events optics and electricity and magneticism appeared as expressions of a single physical mechanism: the transmission of electro-magnetic waves of different lengths in an absolute continuous medium termed the ether. This and all previous knowledge concerning optical, electrical, and magnetic phenomena was summed up in Maxwell's electro-magnetic equations.

Even in this tremendous synthesis certain phenomena remained outside. Maxwell's equations did not apply to the electro-magnetic properties of bodies in motion. The Dutch physicist, H. A. Lorentz, generalized them to bridge this gap. The result was the discovery of the atomicity of electricity. The electron became the elemental entity of nature.[3]

It followed from Maxwell's equations, even before they were generalized by Lorentz, that if a certain experiment were performed it would be possible to determine the velocity of the earth relatively to the static absolute ether. This capacity to determine whether the earth is at rest or in motion by means of an experiment on or within it meant that the principle of relativity did not hold for electro-magnetics. This consequence also exhibited itself in the fact that its laws for a given phenomenon took on a different and simpler form if observations were referred to the ether, than they did with any other Galilean or non-Galilean frame of reference.

The experiment which was to confirm this conclusion and reveal the velocity of our earth relatively to the ether was performed by Michelson and Morley. Remember that it followed as a necessary consequence of Maxwell's equations that this experiment should give a positive result. Nevertheless, the findings were negative.

Consider the situation in which scientists found themselves. The equations of Maxwell, which have behind them an amount of exact knowledge and experimental verification so great as to make their rejection impossible, necessitated an experimental finding which does not exist. The only alternative was to modify the equations of Maxwell and Lorentz in such a way that they hold for traditional knowledge and give rise to the negative result of the Michelson-Morley experiment also. But how was this to be done?

The first attempt at a solution of the problem was offered independently by Fitzgerald and Lorentz. It is known as the Lorentz-Fitzgerald Contraction Hypothesis. It specifies that a body, as it moves through the ether, contracts in the direction of its motion according to a certain formula. This formula is so defined that it introduces precisely the modification which is necessary to adjust Maxwell's and Lorentz's equations to the negative result of the Michelson-Morley experiment. The formula is that the original length of the body is to its contracted length as

$$\frac{1}{\sqrt{1 - \frac{v^2}{c^2}}}$$

where v is the velocity of the body relatively to the Galilean frame from which its velocity is measured and c is the velocity of light. Since the square of c (186,000 miles per sec.) is a velocity so large compared to any velocity with which we are ordinarily concerned, this contraction does not exhibit itself in ordinary observations or experience. Only a refined observation such as the Michelson-Morley experiment could detect it.

It is to be noted that this met the immediate difficulties. One could accept the result of the Michelson-Morley experiment, admit Maxwell's equations and the existence of the ether, and explain our failure to detect our motion through the ether because nature capriciously produces a contraction which exactly compensates for the effect which the ether had been expected to reveal.

But on further consideration, this hypothesis did not appear to be quite so satisfactory. In the first place, it required that all objects regardless of whether they were made of rubber or steel should contract to precisely the same extent. Secondly, it followed, if such a contraction occurred in certain materials, that after-effects in the form of strains should appear. Experiments did not reveal these strains. Thirdly, it was pointed out that there must be a contraction in time as well as in space. This led to a peculiar doctrine of "local times." And finally, it followed that a more refined experiment should be able to go beyond the contraction which nature has introduced and detect a residual motion which would reveal our velocity relatively to the ether.

At this point the brilliant French mathematician of the last century, Poincaré, made a suggestion. He indicated that this was a rather dubious type of scientific procedure.

Man had a theory to the effect that he should find an ether, and when experiment gave the lie to this deduction he still retained his theory by patching it up with hypotheses and devising new experiments to confirm what previous experiment had denied.

This led Lorentz to a very unusual conclusion. He decided to retain all the traditional ideas and put Maxwell's equations into such a form that no possible more refined experiment of any kind whatever should be able to detect the velocity of our earth relatively to the ether. Note what the last part of this involves. It means in practice that the principle of relativity for Galilean frames of reference holds for electro-magnetics as well as for mechanics. For if no possible experiment on the earth can detect whether it is moving or at rest relatively to the ether then electro-magnetics joins with mechanics on the doctrine that there is no privileged Galilean frame of reference.

But Lorentz failed to see the full implications of this point. For he still retained his faith in the ether. This was one of the most peculiar conclusions in the whole history of science. It amounted to an official announcement upon the part of science that nature has entered into a conspiracy against the scientist which is so thoroughgoing that no matter how ingeniously man may try to find the ether which exists, compensations, like the Lorentz-Fitzgerald contraction, are introduced so that it is not discovered. The principle which justified this remarkable anthropomorphic pronouncement bears the harmless formal title, the principle of correlation.[5]

When Maxwell's equations were revised so that this principle held for them, Lorentz possessed several equations termed the Lorentzian transformation equations which enable one, who knows the time and space values for a given phenomenon measured relatively to one Galilean frame of reference, to determine the values for a different Galilean frame, when the velocity of the second frame relatively to the first is known. Also, a certain constant c, which stands for the velocity of light, appeared in these transformation equations, indicating that the velocity of light is the same for any Galilean frame of reference. It remained for Einstein to discover the consequences and real meaning of this jumble of fact and fiction. This brings us to the special theory of relativity.

THE SPECIAL THEORY OF RELATIVITY

When Einstein examined the electro-dynamic equations of Maxwell, at the opening of this century, his attention was called to a certain assymetry in their form which did not correspond to anything in nature. This lack of correspondence between mathematical theory and actual fact first exhibited itself in such ordinary phenomena as the relation of a magnet to a conductor. This assymmetrical character of the equations bespoke the fact that the principle of relativity did not hold for electro-magnetics. Its laws had one form if one referred observations to the magnet and a different form if observations of the same phenomenon were referred to the conductor.

These considerations suggested to him "that the phenomena of electro-dynamics as well as of mechanics possess no properties corresponding to the idea of absolute rest" and that "The same laws of electro-dynamics and optics will be valid for all frames of reference for which the equations of mechanics hold good." [8] In other words, the principle of relativity for Galilean frames of reference applies in electro-magnetics as well as mechanics.

He then assumes this principle as a postulate and adds as a second postulate the principle known as the principle of the absolute velocity of light in vacuo, which affirms that "light is always propagated in empty space with a definite velocity c which is independent of the state of motion of the emitting body." By appealing to a new analysis of the nature of time for simultaneous and spatially separated events, he is then able to deduce the principle of the relativity of simultaneity. These three principles: (1) the principle of relativity for Galilean frames of reference; (2) the principle of the absolute velocity of light in vacuo, and (3) the principle of the relativity of simultaneity constitute the special theory of relativity. The result is an electro-magnetic theory which escapes the asymmetry of the traditional equations, and which reconciles the Michelson-Morley experiment with traditional knowledge by indicating that its failure to detect the absolute ether is due, not to the capriciousness of nature, but to the fact that such an absolute entity, independent of matter, does not exist.

Such was the procedure which Einstein followed in his original paper on this theory, in which his purpose was to present the theory from the deductive point of view. There is another way of presenting it which he followed on another occasion.[6] Since it is more inductive in the sense of indicating how the theory of relativity is a necessary consequence of previously established conceptions, we shall follow it here. For it is very important that the reader senses this theory, not as an expression of two principles set up arbitrarily and hit upon by chance, but as a necessary consequence of previous evidence.

Let us begin with the electro-dynamical theory of Lorentz after his principle of correlation had adjusted Maxwell's equations to the negative result of any possible Michelson-Morley experiment. An examination of these equations would have revealed two principles: (1) the principle of relativity for Galilean frames of reference, and (2) the principle of the absolute finite velocity of light in vacuo. Einstein noted that these two principles contradict each other, on the basis of traditional and accepted conceptions. The first suggestion was that one of the principles must be rejected. But an examination of the evidence revealed that this is impossible. Thus science found itself in the same type of situation, at the opening of our century, that it had faced in connection with the extensive facts of stuff and change at the very outset of its development in Ancient Greece. Two facts, or inescapable verified principles, existed, which contradicted each other.

Einstein pursued the only course which is possible in such a circumstance. He argued that this proves that the presupposition which makes the two verified principles contradictory, must be false, and must be replaced by its negate, which permits the acceptance of the two principles in question, without contradiction.

He proceeded, therefore, to examine traditional theory and note what presupposition makes the two principles contradict each other. He discovered that it is the doctrine of the addition and subtraction of velocities, which rests upon the principle that time is absolute. Nothing remained but to reject the latter principle and regard time as relative. Thus a contradiction in traditional electro-magnetic theory drove Einstein to the discovery of the principle of the relativity of simultaneity. This is the essential contribution of the special theory of relativity.

Just as Parmenides' proof of a contradiction between the extensive facts of stuff and change drove Leucippos and Democritos to the introduction of a referent for atomicity and motion and the formulation of the kinetic atomic theory, so a contradiction in electro-dynamics between the principle of relativity for Galilean frames of reference, and the principle of the absolute finite velocity of light, in vacuo, drove Einstein to the discovery of the relativity of time and the formulation of the special theory of relativity. It is to be noted that in both cases no new evidence is introduced and no experiments are performed. The conclusions owe their validity solely to logic and traditional evidence. They are necessary consequences of established ideas.

The meaning of the principle of the relativity of simultaneity is easy to grasp. According to the absolute theory of time of classical science, nature was conceived as a one-dimensional series of instants. Let us represent this series in our imagination as an endless series of parallel non-intersecting lines. The absolute theory of time means that there is only one such series in nature. Hence if an observer on a given frame of reference makes no error in his clock readings and assigns two events to the same instant in this series, all other observers on other frames of reference who use standard clocks and read without error will make the same assignment. The relativity of time means, however, that the relation of simultaneity between spatially separated events is dependent upon one's reference system. Hence if an observer on one frame correctly observes two spatially separate events to happen at the same time, it does not follow that an observer on another frame who reads his clock correctly will do so. In other words two events may be simultaneous for one observer and not simultaneous for another. From this it follows that there must be a large number of different series of instants and not merely the one absolute series of classical science. In this fashion the special theory of relativity interprets the "local times" which appeared in the classical theory of Lorentz.

It remained for Einstein to determine whether there is anything in the concept of time as it is used in physics which prevents this from being the case. Investigation revealed that the time of physics must be relative, as his deductions had indicated. For the only observation of simultaneity which is intuitively or immediately given in the time determinations of the physicist is that of spatially coexistent events. Only when two events are side by side can we observe their simultaneous occurrence immediately. In other cases we have to appeal to physical motion, or propagation.

Let us take an illustration which Einstein has used.[6] Imagine a level railroad track and suppose that two bolts of lightning strike this track at points two miles apart. We are dealing with spatially separated events. Now, what does the physicist mean by the statement that these two events are simultaneous? Obviously, we do not mean that they are immediately given to the observer as simultaneous, for this will not be the case unless the observer is looking at both at the time of their occurrence and is equidistant from them. This consideration indicates our meaning. Two spatially separated events are simultaneous when the light rays from both leaving upon their occurrence, reach a point equidistant from both at the same time. In other words, the simultaneity of spatially separated events is defined in terms of the simultaneity of spatially coexistent events, and motion.

This point is very important for several reasons. First, it indicates that time is not an elemental concept, but must be defined in terms of motion. The full significance of this has not been appreciated either by physicists or philosophers. Second, it indicates that the theory of relativity is a physical theory. For it is only if one conceives of nature as a physical system containing both physical objects called reference frames, and rods, and clocks, and real motion that the basis exists for the type of temporal relativity which Einstein's theory introduces. Psychologically or phenomenologically it is quite possible to begin one's observation of nature with the simultaneity of spatially separated events, as Whitehead has done.[22] Certainly we see the yellow patch in the sky called the sun over there while we sit here. The sun there is one event, our presence here is another, and they are spatially separated. It is only if one regards nature as a physical system according to which the actual coexistence of two lightning flashes and railroad track are distinguished from the apparent association of these flashes, because of the assumption of an actual moving wave, that Einstein's doctrine of time holds. This has immediate philosophical consequences which are fatal to the position of Eddington, and which present grave difficulties for the philosophy of Whitehead. The former accepts Einstein's definition of the relativity of time and then maintains that relativity physics implies the invalidity of the physical theory of nature.[21] This is a contradiction in terms. For we have just noted that Einstein's definition of the relativity of time does not hold unless one accepts the physical theory.

Also, this point is fatal for all positivistic interpretations of Einstein's relativity theory. For if one begins with phenomena and refuses to accept physical categories then Einstein's theory of relativity cannot be accepted. One must go with Whitehead and admit an immediately given meaning for the simultaneity of spatially separated events. In the latter case difficulties appear. This is not the theory of relativity which physics uses. The theories of Einstein, De Sitter, Weyl, and Eddington rest upon Einstein's definition of simultaneity. Secondly, the whole theory of measuring, both past and present, would take on a purely arbitrary character for which theory would give no justification were this the case. For if space-time theory is a relation between phenomenal events, if the whole of science is to be restricted to sense qualities and phenomenal characters, then the necessity of attaching reference frames to molar bodies and using physical things like clocks and measuring rods, and of drawing a distinction between the special and general theory which turns upon uniform and accelerated motion of physical bodies, is a mystery. Certainly, if science finds space-time to be a relation between phenomenal events, then the crack of a revolver or the smell of a rose should serve as a suitable zero-point for a coördinate system. Needless to say neither pre- nor post-relativity physics reveals any such thing. The reference bodies for natural measurement have been and remain physical objects. And in the doctrine of time as Einstein formulated it, the physical theory of nature and motion is presupposed.

Einstein's definition of time in terms of motion is important also, because when combined with the principle of relativity for Galilean frames, and the principle of the finite absolute velocity of light, the principle of the relativity of simultaneity follows necessarily. To make this clear let us return in imagination to our railroad track and the two flashes of lightning. In order to determine whether the flashes in question are simultaneous, we assume the velocity of light to be absolute, as one of our postulates permits, and establish a point mid-way between the two points at which the flashes occur. We will assume that the two flashes arrive simultaneously at the mid-point M. Hence, our definition of time permits us to assert that the two flashes are simultaneous for an observer situated at M.

It can be shown, if this is the case, that these two flashes will not be simultaneous for an observer moving along the track with a uniform velocity relatively to M. Let us assume that the observer on the uniformly moving train is at the point M when the two flashes, as judged from the ground, strike the railroad track. The principle of relativity for Galilean frames of reference permits him to regard his frame as at rest. The definition for simultaneity which we have assumed also enables him to say that if light rays cover equal distances and fail to arrive simultaneously, then the events which are their source are not simultaneous. It follows, therefore, if the man on the ground judges the events to be simultaneous that the man on the train will not. This becomes clear if one imagines the train to be very long and its two ends to be so close to the ground that the two bolts of lightning strike them in striking the ground. Then the man in the middle of the train will be equidistant from the two events. Since the rays travel from both ends of his train to the middle they travel equal distances, as they do for the man on the ground who is mid-way between the places of the flashes. Obviously, then, if the light rays arrive simultaneously for the man on the ground, they will not do so for the man on the train since the latter is moving toward the ray coming from the front of the train and away from the one coming from the rear. Hence, the former will judge that the events were simultaneous, the latter that they were not.

The fact that such a commonplace occurrence could yield such an unexpected result is somewhat confusing at first sight. It is to be noted that the validity of the proof rests upon three things: the definition for the time of spatially separated events in terms of physical motion or propagation and immediately given coexistent events, the assumption that the velocity of light is finite and constant and the same for either or any frame, and the principle of relativity for Galilean frames of reference. It is the first point which is the key to the final thesis, and the last principle which corrects the psychological tendency that makes the proof appear to be unnatural. The reader is inclined to correct for the velocity of the train. It is to be noted that this involves the assumption of a privileged or absolute frame of reference and is contrary to the principle of relativity for Galilean frames of reference. It is to be noted also that time would be absolute even upon this basis for both observers were the velocity of light infinite rather than finite.

It is essential that the meaning and force of the principle of the relativity of simultaneity be grasped. For it is the essential contribution of the special theory of relativity. As we have indicated, both of the other principles which go to make up the theory existed before Einstein in classical electro-magnetics and mechanics. It is the discovery of the relativity of simultaneity which is his unique contribution, and which causes the Michelson-Morley experiment and the principles of relativity and the absolute velocity of light to take on the new interpretation and to give rise to the new deductions which the special theory of relativity entails.

These new interpretations and consequences are many in number.[3] First, as we have noted, the absolute ether independent of matter does not exist. This follows from the denial of a privileged frame of reference, which the principle of relativity introduces. Second, absolute space does not exist. This can be deduced directly from the relativity of simultaneity. A length, measured in a frame relatively to which it is at rest, gives different values than in the case of a frame relatively to which it is in motion. The Lorentzian transformation equations tell what this difference is. It happens that it corresponds exactly to the Lorentz-Fitzgerald contraction. Thus the correction, which Lorentz and Fitzgerald introduced, exists, but it has its basis not in the actual metaphysical contraction of the body, but in the difference of measured values when referred to different Galilean frames of reference. In this way the failure of experiment to reveal the strains which the contraction hypothesis predicted is explained. Note how beautifully the theory of relativity fits in with fact.

Third, the relativity of measured mass and the hypothetical equivalence of mass and energy follow. These are consequences of the special theory which the reader will do well to check with original sources,[7] and the premises which justify them. There are reasons for believing that they have been wrongly interpreted by certain writers.

Finally, the most essential contribution of the special theory is the union of space and time. In fact the theory can be summed up in this one statement. Of course, this is but another way of saying that the principle of the relativity of simultaneity is the key to the theory. For if there is no meaning to time except as one specifies one's location, and if the relativity of time entails the relativity of space then space and time are, as Minkowski said, inescapably bound together. This point was expressed mathematically by the latter when he developed a four dimensional space-time geometry in which the temporal dimension is given the negative and rather complicated value which makes it symmetrical with the three spatial dimensions.[7] This symmetry is essential to express the fact that the four dimensions of space and time can be assigned between space and time by observers upon different frames of reference in different ways. Thus what is a temporal dimension for one observer on one frame may appear as space for another on a different frame. Since the four-dimensional manifold is Euclidean or semi-Euclidean and never changes its metrical uniformity or properties, it may be said that the space-time of the special theory is absolute. This is the justification for our earlier statement that the special theory of relativity replaced the absolutes, space, time, and the ether, with another absolute termed space-time.

But the most important consequence of the special theory of relativity is that it reveals a gapping inadequacy which forced Einstein on to the discovery of the general theory. The simplicity which it introduced into traditional theory was remarkable. The Michelson-Morley experiment was reconciled with the traditional electrodynamical equations without assigning anthropomorphic attributes to nature; contradictions were removed from traditional theory with the rejection of a large number of concepts and hypotheses by merely allowing the existing contradiction to force one to its logical consequences; and the principle of relativity for Galilean frames of reference was put upon the same footing in electro-magnetics that it enjoyed in mechanics. But relativity still attached to one's laws of nature. For remember, in the special theory of relativity, as in Newtonian mechanics, the principle of relativity which removes the dependence of the form of one's laws upon reference frames, only holds for Galilean frames of reference. Laws of a different form would be necessary if measurements were referred to non-Galilean frames. Certainly, this constitutes an inadequacy in our laws. Nature goes on regardless of the frames we choose, or whether we choose any frames. Laws which purport to apply to nature should exhibit the same absoluteness and independence. Does not our success in attaining this result for Galilean frames suggest that we can do it for non-Galilean frames as well? The answer to this question is the general theory of relativity.

THE GENERAL THEORY OF RELATIVITY

The theme of relativity physics is motion. The special theory arose when absolute motion or propagation was presupposed in defining the simultaneity of spatially separated events. Only by assuming a motion or propagation which is absolute, since it is not defined relatively to coördinate systems, can one get the special theory of Einstein. This is even more obvious with respect to the general theory. Its aim is to extend the principle of relativity so that it applies to non-Galilean or accelerated, as well as Galilean or inertial frames of reference. The whole point in introducing this principle is to escape the relativity which a use of coördinate systems in measuring entails. Science wants its laws in such a form that they refer to nature itself and not to an arbitrarily chosen relation of nature to a reference body. In short, the generalization of the principle of relativity is the expression of a quest for the absolute. Now, what is this absolute which the general theory is trying to disentangle from our traditional arbitrary formulation in terms of arbitrarily chosen coördinate systems? The subject matter of the science to which the general theory applies gives the answer to this question. This science is mechanics, and its subject-matter is the motion of masses. Hence, it is the motion of masses which the general theory of relativity succeeds in expressing in absolute terms. Thus the general theory, like the special, presupposes absolute motion. To interpret it in any other way is to forget the subject-matter of the science to which it refers, and to miss the whole point in extending the principle of relativity.

We have noted how the general theory grows naturally out of Einstein's previous work. The special theory had gone but part way in enabling our laws of nature to transcend the relativity which measuring and its reference frames introduces. The principle of relativity had been extended from mechanics to electro-magnetics but it held in both fields only for observations made from Galilean frames of reference. Is it not possible to put our laws of mechanics into such a form that they will apply to masses and motions themselves and hold for any frame of reference whatever?

The philosophical desirability of such a formulation of the laws of motion is evident. But does nature permit it?

At this point two considerations guided Einstein.[9]

First, Mach had revealed an inherent epistemological defect in Newtonian mechanics which the acceptance of an unrestricted principle of relativity removes. Second, because of the failure to extend this principle to non-Galilean frames, Newtonian mechanics imposed certain restrictions upon nature which nature does not exhibit. Einstein puts the matter somewhat as follows, in his original paper on the general theory. Consider three masses *a, b,* and *c.* We have added a third mass to Einstein's illustration.[9] Let *a* be sufficiently far from other masses to be unaffected by external forces, let *b* move with a uniform velocity or remain at rest relatively to *a,* and *c* move with a uniformly accelerated motion relatively to *a* and *b.* According to Newtonian mechanics the laws of motion hold only if one observes and measures the object *a* relatively to the reference frame *b,* for *a* is not acted upon by external forces and hence, obeys Newton's first law of motion only for observations made from mass *b.* Since *c* is accelerated relatively to *b* and *a,* the objects *a* and *b* will have an accelerated motion, relatively to *c,* the direction and amount of which is independent of the "material composition and physical state" of the masses in question. Now this is precisely the type of motion which a gravitational field produces. Thus it becomes evident that nature does not restrict one, as does Newtonian mechanics, to the statement of *a*'s motion relatively to the Galilean mass *b.* It gives just as good justification for the use of mass *c* as a reference frame, in which case the observed and measurable acceleration of *a* can be interpreted as due to the presence of a field of gravitation around *a.* At least this is true so far as qualitative considerations provide a criterion for judgment.

But do the observed masses themselves when considered quantitatively permit this? It is to be noted that when we consider the body *a* relatively to the frame *b,* we are regarding it as an inertial mass, whereas when we are considering it relatively to the non-Galilean frame *c* it appears as a gravitational mass. Can we shuffle these two masses of any body in this fashion? When the question was put in this form it became evident that it is not necessary to look afresh at nature to determine whether one can extend the principle of relativity to all possible physical frames of reference; in the principle of the equivalence of inertial and gravitational mass classical science itself contains the justification for this extension. For the latter principle expresses the verified fact that these two masses, which Newtonian mechanics had regarded as essentially different, are numerically proportional and the same; in other words, the inertial mass and the gravitational mass of a body are not two different things, but two aspects of the same thing. Thus it became evident that classical science provides the justification for regarding the same mass as an inertial system relatively to one frame and an accelerated or gravitational mass relatively to another. In this fashion the verified classical principle of the equivalence of inertial and gravitational mass was given theoretical justification for the first time and became the foundation of the general theory of relativity. Thus verified fact joined with logical consistency and the scientific quest for objectivity to warrant and urge the acceptance of the unrestricted principle of relativity. The laws of the motion of masses must hold for observations and exact space and time measurements made from any physical frame of reference whatever. This is the general theory of relativity.

Note what it entails. Since the motion of the object *a* can be referred to the body *b*, in which case no gravitational field exhibits itself about *a*, or to the body *c* in which instance *a* appears to be accelerated by a gravitational field, it follows that gravitation is relative. It can be brought into or put out of existence in certain cases by a suitable shift of coördinates. Hence, Einstein knew at the very outset that the unrestricted principle of relativity necessitates entirely new laws of motion and a new law of gravitation, since the traditional laws were stated in terms of absolute gravitation.

What can these new laws be? A second consequence of the unrestricted principle of relativity provides the answer to this question. In Newtonian mechanics space and time had objective physical significance. The special theory had revealed both to be relative, but meaning still remained for an objective absolute geometry, or chrono-geometry, which was Euclidean, since space-time had been substituted for space and time, and no possible choice of permissible frames of reference could remove its invariant Euclidean properties. With the acceptance of the unrestricted principle of relativity, "the last remnant of physical objectivity," to use Einstein's words, "is taken away from space and time." [9] Not merely space and time but space-time and geometry itself is revealed to be completely relative. The traditional procedure of describing natural processes in terms of a relation of matter to a geometrical entity such as space or space-time is an error. A given geometry or chrono-geometry whether it be Euclidean or Riemannian, like a gravitational field, can be brought into or put out of existence by a suitable choice of reference body. In other words, the geometrical continuum which we supposed, both in Newtonian mechanics and the special theory of relativity, to be the background for the occurrence of natural processes is not objective at all, but has its basis solely in the relation of moving masses to the particular physical object which we pick out as a reference body. Not merely the space and time readings but the metrical or geometrical rules themselves, which we use to coördinate and relate those readings, are relative.

This discovery is very important for it confirms a conclusion reached before Einstein by mathematical logicians and students of pure mathematics who analysed the nature of mathematics. In fact, Einstein says that their work influenced him in the discovery of the general theory.[12] Their investigations revealed that there are no mathematical entities. The subject-matter of mathematics is not an objective thing. In fact, mathematics is not talking about anything in particular. As Bertrand Russell has said, "Mathematics may be defined as the subject in which we never know what we are talking about, nor whether what we are saying is true." This statement is literally exact. It means that mathematics asserts implications between expressions that contain variables. Now, variables are symbols which refer to specific individual values but not to a specific one in particular.[26] Hence it is not until one substitutes values given in an empirical science, for the variables with which mathematics concerns itself that one can tell what one is talking about. This does not happen until one gets outside of mathematics. The point is that the implications which mathematics asserts hold for any entities whatever; hence there is no point, so far as mathematics is concerned, in asserting them for only a specific subject-matter. It becomes evident therefore that it is as ridiculous to conceive of our universe in terms of the relation of matter to a geometrical continuum, as it would be to conceive of it in terms of the relation of matter to the syllogism. For a continuum is not an extensive entity which can serve as the necessary referent for atomicity and motion, but is a set of implications. Mathematics is a chapter in formal logic rather than physics.[25] It is Einstein's great merit to have sensed the physical significance of these discoveries of the mathematical logicians and to have revealed, in the general theory of relativity, that there is empirical as well as analytical and formal evidence for them.

It follows automatically of course from the discovery that geometry or chrono-geometry possess no objective existence as the background for physical processes, that a new non-mathematical entity must be introduced to provide the necessary referent for atomicity and motion. It happens, however, that Einstein, probably because of his failure to recall the established conclusions of Greek science, overlooked this point. In fact, we shall find that most of the physicists since Einstein, and to some extent even Einstein himself, have made relativity theory a muddle of contradictions, by taking space or space-time, which the general theory reveals to be relative, and smuggling it in again as a referent for motion or as the guiding field for matter. So great is the tendency of scientists to worship theories expressed in mathematical terms, rather than to understand either their meaning or the nature of mathematics. This tendency has been particularly vicious in connection with the theory of relativity because it has caused practically every scientist who has given an exposition of it, except Einstein and Whitehead, to write like a degenerate Hindu mystic rather than a physicist. Consequently, in the name of science, we have had to read a lot of nonsense about our universe having another dimension, and matter and its motion being determined by puckers in space-time. Such statements, which are commonplace in current scientific literature, reveal that their authors have not grasped one of the most elemental points expressed quite clearly and unequivocally in Einstein's original paper on the general theory.

This point with its obvious implications has been overlooked so often that it is well to have Einstein's own words concerning it. In his original paper on the general theory [9] he writes "In classical mechanics, as well as in the special theory of relativity, the coördinates of space and time have a direct physical meaning. . . . This view of space and time has always been in the minds of physicists, even if, as a rule, they have been unconscious of it. . . . But we shall now show that we must put it aside and replace it by a more general view, in order to be able to carry through the postulate of general relativity . . ." He does this by demonstrating with reference to a flat circular disc rotating about its central point that the ratio between the lengths of the diameter and radius of the circle which it involves, obeys only Euclidean rules if measurements are made in the Galilean frame which is a plane at rest on which the circumference and radius of the rotating disc is marked, whereas the ratio will obey only the rules of Riemannian geometry if measurements are referred to the rotating disc itself, which is a non-

Galilean frame. This follows from the special theory of relativity because viewed relatively to the Galilean frame the measuring rod undergoes a contraction because of its motion in measuring the circumference which it does not undergo in measuring the radius of the rotating circle, whereas no contraction would appear in similar measurements on the frame which is at rest. "Hence," Einstein continues, "Euclidean geometry does not apply to (the non-Galilean frame) K'. The notion of coördinates defined above, which presupposes the validity of Euclidean geometry, therefore breaks down in relation to the system K'."[9] He could also have added, what many seem to have overlooked, that the Riemannian metric of frame K' has no better status on frame K. All geometry is relative. It is to be noted also that it is merely to facilitate exposition, that he speaks of geometry rather than of chronogeometry. Space-time whether Riemannian or Euclidean has no better status than space. We can understand therefore why Einstein says on the next page "That this requirement of general co-variance, . . . takes away from space and time the last remnant of physical objectivity," and why in a later paper[10] he writes that "In a consistent theory of relativity there can be no inertia *relatively to 'space,'* but only an inertia of masses *relatively to one another.*" The empirical evidence which urges the acceptance of the unrestricted principle of relatively necessitates the conclusion that there are no objective geometrical entities of space or of space-time, whether Euclidean or Riemannian. The general theory of relativity has not replaced the Euclidean space of traditional science and common sense, or the Euclidean space-time of the special theory with Riemannian space-time. Instead it has revealed that all such systems of relations are relative since they have their basis solely in the relation of the absolute masses and motion which is the subject-matter of mechanics to the particular mass which one arbitrarily chooses as a reference body for measuring.

It is to be noted therefore, that the general theory of

relativity is a more thorough-going physical theory than any which science has previously known. Whereas the traditional physical theory of nature introduced a geometrical entity termed space as the referent for atomicity and motion, this theory divorces from all spatial, geometrical, or chrono-geometrical entities the last vestige of physical objectivity and independent meaning. We shall find that the full consequences of this fact have not been recognized.

But we must return to Einstein's problem. The acceptance of the unrestricted principle of relativity forced him to the conclusion that gravitational fields and geometries are relative. Both can be put out of existence in certain instances by a suitable shift of coördinates. Since the traditional laws for mechanics were stated in terms of absolute gravitation and absolute Euclidean space or space-time, the necessity of a new formulation of them becomes more and more evident. But how is such a new formulation to be made? We undertook to extend the principle of relativity in order to lay hold of the absolute motion of physical objects themselves, but instead of getting nearer to the absolute we seem to find ourselves with more and more that is relative. For when we describe the motion of a given physical object from a Galilean frame of reference we find no gravitational field present and must use a Euclidean metric, whereas when we describe the same moving object from a non-Galilean frame a gravitational field springs into existence and a Riemannian space-time metric is necessary. Obviously if Einstein had stopped at this point the laws of mechanics would have had one form for one frame and an entirely different form for another, and the whole point in accepting the principle of relativity would have been lost. Furthermore our laws would be stated in terms of a geometry, and geometry we have found to be relative.

Certainly if the principle of relativity is valid, there must be some form for the laws of the motion of physical objects in this universe which will remain constant through all these different relative gravitational and geometrical descriptions. This is what the principle of relativity means; the laws of a given natural phenomenon possess precisely the same form regardless of whether the phenomenon is measured from one frame of reference or another. Obviously such a law cannot be stated in terms of space-time or geometry since geometry is relative; it must be a law with a changeless form, the values of which in a given frame, defines the presence or absence of gravitation and the specific space-time metric for that frame. Now, is there a logical form containing variables the values of which define a geometry or chrono-geometry? Einstein went to the scientists, whose business it is to discover all possible logical forms, for an answer to this question. They are usually called pure mathematicians. It would be less misleading and there would be less chance of the false worshipful attitude toward mathematics arising, if they were called mathematical logicians. They informed him that such forms exist. They are known as tensors.[5] They also informed him that there are a large number of such tensors, and queried him concerning his demands, in order to determine the one which physics requires. Upon being told that the values of its variables must define the metric of a given frame of reference the answer was easy. For mathematics indicates that the tensor, the values of the variables of which, will define a relative space-time and remain invariant in form for all possible transformations, or in other words for any shift of reference frame, is determined completely by the number of dimensions of the metric which appears in a given frame. Furthermore the tensors for the different dimensional relative space-times are defined by the number of variables, termed g_{ik}'s, which they contain. A symmetrical tensor whose variables take on values in a given frame which define a relative two dimensional metrical field, contains three g_{ik}'s; one for a three dimensional field, six g_{ik}'s; and one for a four dimensional field, ten g_{ik}'s. Since the special theory has indicated the latter to be the case Einstein knew that an adequate law for the motion of masses must be a tensor form containing ten g_{ik}'s. This particular tensor follows: [8]

$$ds^2 = g_{11}\,du_1^2 + 2\,g_{12}\,du_1\,du_2 + 2\,g_{13}\,du_1\,du_3 + 2\,g_{14}\,du_1\,du_4 + g_{22}\,du_2^2 + 2\,g_{23}\,du_2\,du_3 + 2\,g_{24}\,du_2\,du_4 + g_{33}\,du_3^2 + 2\,g_{34}\,du_3\,du_4 + g_{44}\,du_4^2$$

where du_1, du_2, du_3, and du_4 represent the differences between two points infinitely close together, in each of the four Gaussian coördinates, and the numbers which prescribe the relations between these values at a given point are designated by the ten g's, — g_{11}, g_{12}, g_{13} etc., termed for short, g_{ik}'s.

This is summarized to read: [8]

$$ds^2 = \Sigma g_{ik}\,du_i\,du_k.$$

Mathematics also informed Einstein that when these ten g_{ik}'s take on the specific values, in a given frame, of O, –1, and 1, in a certain fashion, the frame in question is Galilean and its metric is Euclidean; [9] when any other values appear the metric is Riemannian and the frame is non-Galilean. Thus the demands of the unrestricted principle of relativity were realized. For this tensor possesses a form which remains invariant for all possible frames of reference, and nevertheless necessitates that different geometries must appear in the different frames.

But one other important discovery was also at hand. We have noted that a shift from a Galilean to a non-Galilean frame of reference entails, not merely the replacing of a Euclidean with a Riemannian metric, but also the introduction of a gravitational field where there was none before. The conclusion is obvious. Gravitational potentials and the g_{ik}'s which define a Riemannian metric are identical. Riemannianism exhibits itself in the distribution of potentials of the gravitational field which goes with our frame. It was this discovery of the identity between the gravitational potentials of the physicist and the values of the g_{ik}'s in a non-Galilean frame which is one of Einstein's greatest achievements in the general theory. Riemannianism is not something foreign to our experience holding only for some other hypothetical world. Notwithstanding its relativity, it is all around us since we are on a non-Galilean frame. The gravitational field which accompanies our frame is an immediate evidence of it. We can now understand why Einstein wrote: "Thus, according to the general theory of relativity, gravitation occupies an exceptional position with regard to other forces, particularly the electro-magnetic forces, since the ten functions which define the gravitational field at the same time define the metrical properties of the space measured."[9] Such is the importance of the discovery that gravitational potentials and the values of the g_{ik}'s which define a relative Riemannian metric are identical.

This discovery is important for other reasons. First, since an invariant law for mechanics can only be stated in terms of a tensor containing ten g_{ik}'s Einstein knew that an adequate law for gravitation must contain ten potentials instead of the one in Newton's law. In this fashion the science of pure forms contributed to the science of experimental physics. Investigation revealed that the Newtonian gravitational potential is the variable g_{44} in the tensor given above. Second, it enabled one to define the range of application of Newtonian mechanics and the special theory. Strictly speaking no mass is so completely isolated from other masses that no gravitational field appears relatively to it. Nevertheless we know that Newtonian mechanics is approximately true, and uses Galilean frames and Euclidean geometry. It was assumed therefore that the fixed stars approximate so nearly to rest in Euclidean space, that the Newtonian procedure applies in most cases. We shall see that this thesis was somewhat unfortunate in its results because it led some to deny the general theory by conceiving by our universe again as a collection of physical objects in Euclidean space. Thus the doctrine of the relativity of geometry which is essential to the general theory was denied. Instead, Galilean frames and their Euclidean metric should be regarded as a limit which only exists in an ideal sense. This means of course that actually there are only non-Galilean frames. Since they always entail a Riemannian metric it may seem as if the general theory in the end, restores a Riemannian metric to an absolute status. A slight reflection on the peculiarities of Riemannian geometry as compared with Euclidean, will indicate however that this is not the case. For whereas there is only one Euclidean metric, there are an infinite number of Riemannian metrics depending upon the degree, for example, to which the sum of the internal angles of a triangle is greater than two right angles. Hence to say that the geometry of any frame will be Riemannian is not to assert that there is one Riemannian metric which remains invariant for all frames. It is always the distribution of matter relative to a given frame which determines the gravitational field and hence the degree of curvature of the Riemannian metric of that frame. Thus we make another discovery. Geometry is not merely relative but even the relative geometry of a given frame is determined by matter and its distribution. Thus Einstein writes, "According to the general theory of relativity the metrical character (curvature) of the four-dimensional space-time continuum is defined at every point by the matter at that point and the state of that matter."[10] Not only does the general theory divest space, time, and space-time of all objectivity but it defines even their purely relatively character in physical terms. Never has there been such a thorough-going physical philosophy of science as we find in the verified general theory of relativity. Even space, time, and space-time which have been regarded previously as independent of matter, are discovered to be relations between physical objects.

Nor should this surprise us. For if the relational implications with which mathematics is concerned do not refer to mathematical entities it follows that any application which they have to nature must be defined in terms of the subject-matter of the science in which the mathematics is used. Since the science with which Einstein is concerned is mechanics, and its subject-matter is masses and their motion, it follows that the mathematics of the relativity theory must be defined in physical terms. Einstein has said precisely this in a paper entitled Geometry and Experience[12] in which he indicates that the constants, which are to be substituted in the formal equations of mathematics for the variables of pure mathematics, are solid bodies. This point is important not only because it indicates unequivocally that his theory necessitates the acceptance of the physical rather than the mathematical philosophy of science, but also because it means that we must go to the physics as well as the mathematics of Einstein's theory to find out what it really means.

The additional advances[5] which Einstein had to make before his task was completed reconfirm this conclusion. We have noted that the acceptance of the unrestricted principle of relativity drove him to the statement of the laws of motion in terms of a tensor form containing ten g_{ik}'s. But this tensor only describes a motion of infinitely short extent. The symbol d which appears throughout the tensor expresses this fact. Now, the motions of masses in this universe are not of this character; they are at least finite in extent. Hence an adequate law must be of an invariant form which ties the g_{ik}'s describing one infinitesimal amount of motion to those defining another. Again pure mathematics came to the rescue. A curvature tensor has this formal property. But there are a great number of such tensors: G, G_{ik}, B_{ikst}, and others.[5] Which one applies to the motions of physical objects in this universe? Were Einstein's theory concerned with space-time rather than with masses and motion, it would have had a purely arbitrary character beyond this point. For, as far as pure mathematics is concerned, one of these tensors is as good as another. But the beauty of Einstein's general theory is that the final mathematical tensor equation at which he arrives is unique. It is this character of it which would force men to take it seriously even though it were not verified; it was this uniqueness which caused Einstein to publish his results with faith in their validity before the experiments which confirmed his judgment were performed. Once the unrestricted principle of relativity is accepted there is no stopping until one has progressed far beyond the stage at which we have now arrived. Not merely logic but physics guided Einstein.

At this point it was physics which provided the necessary answer. Newton's law fits nature very exactly and almost, but not quite, universally. It has a mathematical form of the second order. A law fitting this universe cannot deviate very far from Newton's law. Hence a law defining a finite motion must have a tensor form of the second order. This and other physical considerations eliminated the curvature tensor B_{ikst} and centered attention on the tensor G_{ik}, which is of the second order as its indices indicate. But this curvature tensor does not possess properties which insure the conservation of matter.[5] Physics indicated that at least an approximate conservation exists. Another curvature tensor of the second order, very much like this one, meets this requirement. It has the following form: [5]

$$G_{ik} - \frac{1}{2} g_{ik} G.$$

The changeless form of this equation describes the absolute motion of a given mass. But we know not merely that bodies move but that their motion is determined by their relation to other masses, and the physical field in which they are located. An adequate law must contain another expression describing other masses or matter, and this additional expression must be set equal to the one given immediately above to express the control of other masses, or matter in general, over the motion of the specific mass which the tensor given above describes. Furthermore if this additional expression is to satisfy the unrestricted principle of relativity it must be of a tensor form containing ten variables. It happened that Einstein possessed such a tensor describing matter. It is designated by the symbol T_{ik} and is known as the matter-energy tensor. The result follows: [5]

$$G_{ik} - \frac{1}{2} g_{ik} G = T_{ik}.$$

This is the tensor equation of the general theory of relativity. It is fair to say that it comes nearer to that which is absolute in our universe than any equation of science which expresses an approach to measured nature through coördinate systems. It is worth our while therefore, in the light of our previous analysis, to reflect a moment concerning its meaning, and to dwell upon its different aspects. In the last analysis its meaning is perfectly simple. It merely designates how the absolute motion of a given mass is controlled by its absolute relation to the rest of matter. All that Einstein has done is to discover that empirical considerations as well as philosophical ones urge the acceptance of the unrestricted principle of relativity, from which it follows of necessity that Newton's formulation of the laws of motion and of gravitation and the status of geometry is restricted and relative since it refers to an arbitrary relation of the observer to the objective motion, instead of to the objective motion itself; hence, it must be replaced by a law of a different form. Mathematics enters merely as a formal logic to enable one to deduce the correct law from the unrestricted principle of relativity and known physical evidence. It happens that the route which logical necessity follows is somewhat long. Hence the mathematics is not the most simple. It happens also that the relative relations between observed phenomena and a particular reference frame which this mathematical reasoning reveals have the formal properties of certain types of geometry. Hence the mathematics has a geometrical emphasis when one is talking about the relative aspect of the theory. But this relative aspect of the theory which is defined by the values of the G_{ik}'s in a given frame must not be confused with its essence; it is to the pure form of the tensor equation that we must look for its objective meaning and philosophical significance, and this, we have noted, refers not to space-time which is purely relative, but to the subject-matter of mechanics, which is masses and the control which other masses have over their motion.

It is to be remembered that relativity is the price which we have to pay for a statement of the physical theory of nature in terms of the near at hand, and for measurable verifiability of our scientific theories in the local instance. Were we limited to a statement of the laws of nature in terms of the immediately measurable they would be verifiable but completely relative; were we to form them solely by hypothesis, in terms of the absolute, they would refer in meaning to objective nature itself but we could never determine whether they are true or not. The great importance of the principle of relativity, first discovered by Galilei, is that it combines both values; and the great contribution of pure mathematics, with its distinction between the changeless form of its equations and the relative values, which measured observations in a given frame give for its variables, is that it enables science to combine the absolute and the relative in a single expression. Thus verifiable laws which reveal the nature of the objective are obtained.

These two aspects of Einstein's tensor equation must be kept clearly in mind. Otherwise the reader will be misled by his law of motion. Let us note how this law arises. We have discovered that the relative gravitational and geometrical fields of any frame are unified, since the values of the g_{ik}'s determine the potential distribution of the former and the metrical structure of the latter. As the potention distribution of a physical field determines the motion of a body in that field, it follows since it also defines the metric, and, hence, any geodesic or shortest distance, that the general theory enables one to replace Newton's three laws of motion with the single law that a body always moves in a path which is a geodesic in the four-dimensional space-time of the frame in which it is measured.

This law is always true, but two things are to be noted concerning it. First, in the same frame of reference, the metric and geodesic for one distribution of matter will not be what it is for another. Thus it becomes evident that it is not space-time but matter that determines the path of a body. This makes it clear also that there are two senses in the general theory in which space-time is relative, and its relativity is defined in terms of matter: (1) The metrical properties which define it may vary with a constant distribution of matter due to a shift from one physical object to another as a reference body; or (2) they may vary in the same frame of reference due to a redistribution of matter. The general theory provides an objective meaning for the distribution of matter; it is given in the changeless form of the T_{ik} tensor, which, when specified, defines the curvature tensor and determines the specific metric and geodesic that will appear in a given reference frame. Certainly, a theory which defines space-time in terms of matter in this fashion makes the physical theory of nature primary and cannot use space-time to define what it means by the atomicity and motion of that matter. Second, although Einstein's unitary law of motion is always true, it is a statement of the general theory in terms of its relative aspect. Such a statement is necessary from the point of view of verification, since verification entails measurements, which in turn involve reference frames and the different relative geometries that go with them. But this does not mean that matter is located in a background of space-time, or that its motion is determined by the structure of space-time. To reach such a conclusion is to confuse the relative values of the g_{ik}'s with the changeless form of the tensor equation in which the g_{ik}'s occur. As Einstein has said, the motion of matter is controlled only by matter, "there can be no inertia" of bodies "relative to 'space'," or we may add, space-time, "but only an inertia of masses relatively to one another."[10] It is to the changeless form of the tensor equation that we must go for our conception of the basic nature of things and this gives only matter and its motion an objective status.

Once this is grasped it becomes evident that the general theory of relativity, as it came from Einstein, cannot be the end of the story. For our knowledge of the consequences of such a physical theory, consequences established by Greek science, indicates that if nature is physical and contains motion, then it must be kinetic atomic in character, and there must be a referent for atomicity and motion which is other than the moving microscopic particles. As matters now stand, with the traditional referent, absolute space, non-existent, and space-time eliminated, and no geometrical structure permissible, no such referent exists. In other words, the general theory of relativity gives a primacy to the physical theory of nature more thorough-going than ever supposed before, only to leave it in a form which is meaningless. Certainly such is the case when the necessary basis and meaning for *many* physical entities, and motion is not provided. It seems to be evident, therefore, that Einstein's theory can not be complete until a new referent is provided for atomicity and motion. Since geometrical structures are ruled out by physical as well as logical considerations, necessity and consistency seem to demand that this new referent must be a physical entity.

But we are now talking about the necessary consequences of the general theory. It is necessary to consider the more speculative generalizations beyond the general theory before going more deeply into this important subject.

Our present position may be consolidated by a summary of our findings. The general theory is a necessary consequence of the acceptance of the unrestricted principle of relativity. Its final form appears in the tensor equation for gravitation. This equation refers to the ordinary physical objects of our ordinary world and merely specifies how the motion of any one of them is determined by its relation to others. In the last analysis this is all that there is to it. It has appeared to be very mysterious and complex only because its complicated relative mathematical aspects have been accentuated and its essential physical meaning has been ignored. It differs from Newton's law in that it is more exact, since it contains ten potentials rather than one, and that it lays hold of the absolute motion and relations of physical nature itself, rather than the relative motion and aspects which appear in a given reference frame when one measures. This occurs because it refers to the subject-matter of mechanics which is matter and motion and attains a formulation of this subject-matter in a form which remains invariant through all possible relative geometrical descriptions and all changes of one's body of reference. In doing this it reveals that the metric of space-time is relative in two senses and that it is conditioned by matter. Hence, it reconfirms the validity of the physical theory of nature. Nevertheless, it has left this philosophy of science in a form which is meaningless, since no referent exists for atomicity and motion. We must expect, therefore, to find it necessary to amend the traditional kinetic atomic theory by adding a new referent which is a physical entity, when we consider the necessary consequences of the special and general theories.

It happens that the general theory of Einstein takes care of every observed fact which Newton's theory included, accounts for phenomena which were inexplicable on Newtonian principles, and necessitates unexpected consequences which have been verified.[3] Hence to logical and physical simplicity, and beauty of dialectical development from traditional fact to current consequence, we must add the statement, "It is true." To accept it is not merely a pleasure but a necessity.

GENERALIZATIONS BEYOND THE GENERAL THEORY

When we pass beyond the general theory and its necessary consequences we enter upon more uncertain ground.

Assumptions are made which do not possess the certainty of those upon which the special and general theories rest, and experiments have not been performed to check one's findings. Nevertheless there are certain considerations which indicate that generalizations and new applications must be made, and that certain of the proposed theories seem to be the most probable answers that are available. These generalizations fall into two groups. There are the theories of the finite universe of Einstein[10] and De Sitter,[19] and the unitary field theories of Weyl,[11] Eddington,[19] and Einstein.[14]

Theories of The Finite Universe

Let us consider the former first. We have noted how the general theory revealed the identity of the relative gravitational and metrical fields of any given frame. This was expressed by the fact that it is only the values of the g_{ik}'s, or variables in the tensor equation, which define gravitational and geometrical properties, and that the values which define the one field, automatically prescribe the other. This dependence of the gravitational and metrical field upon the relative values of the g_{ik}'s rather than on the changeless form of the tensor law indicated that a suitable shift of coördinates could produce either a Euclidean metric and no gravitational field, or a Riemannian metric of a certain curvature and the presence of a gravitational field of a certain kind. It is to be remembered, however, that the theory is a physical theory concerning this physical universe. Hence, the question as to what geometries or gravitational fields will actually arise in any possible frames of this universe, is a question concerning what kinds of possible physical objects, nature presents to us as reference bodies. Certainly the theory of relativity does not force one to use a Euclidean metric if there are no Galilean frames of reference in this universe. Naturally, this empirical question arose. The reply seems to be obvious. No object is so completely isolated from the rest of matter that it exhibits no gravitational effects, and is a perfect Galilean frame of reference. In other words, strictly speaking all physical objects in this universe are non-Galilean frames of reference.

But if this is the case, how does it happen that Newtonian mechanics, which does not hold unless one's frames are Galilean, is so nearly correct? As we have noted, this question was answered, with the thesis that the fixed stars approximate so nearly to a state of isolation that their non-Galilean character can be neglected, for the most part. But instead of stopping at this point and regarding Galilean frames as an ideal limit having no actuality in nature, certain physicists went on, as Newtonian mechanics had done, to give this ideal notion of matter located in Euclidean space an objective status. This of course was a sheer denial of the general theory of relativity, since it gave to Euclidean space the ontological status which the general theory and modern logic denies to such mathematical structures. Thus, far from matter space was regarded as Euclidean.

1. *Einstein's Cylindrical Universe*

Einstein noted that this was inconsistent with the general theory. We find him writing, "In a consistent theory of relativity there can be no inertia of bodies *relatively to 'space,'* but only an inertia of objects *relative to one another.*"[10] He noted also that the notion of physical objects in an infinite Euclidean space is untenable on physical grounds, which are entirely independent of relativity considerations. The doctrine is as untenable in Newtonian mechanics as in relativity physics. Were it true our universe would be unstable. In fact, it can be shown, if statistical considerations are admitted, that our universe could not have come into existence in its present form. Obviously, something is wrong in both Newtonian mechanics and our interpretation of the general theory concerning our conception of the boundary conditions of the universe.

The solution of this difficulty seems to be fairly obvious.

The notion of physical objects in an infinite geometrical space is wrong. This combination of physical and geometrical entities is a muddle. The physical part of nature is all that there is. It is impossible for any physical object to move off from the rest of matter into empty space never to return. In other words, our universe is purely physical and is finite. The general theory in its original form had prescribed the first of these two conditions, but it did not prescribe the second. Hence an addition or amendment to the tensor law had to be made.

The sound procedure to use in producing this amendment seems to be quite evident. A limiting condition which will prevent any part of finite matter from moving away from the rest, must be introduced into the tensor equation, and, since no objective mathematical conditions are permissible, both the finiteness of matter and this limiting condition must be defined in physical terms.

Unfortunately, notwithstanding the correctness of his intentions as expressed in his theoretical utterances,[10] Einstein failed in actual practice to do this. For he merely added a tensor containing a limiting constant termed λ to the tensor form of the general theory to get the following tensor [5]

$$G_{ik} - \tfrac{1}{2} g_{ik} G - \tfrac{1}{2} \lambda g_{ik} = T_{ik},$$

and then failed to define this mathematical symbol in physical terms. Thus when he states in his original paper on the theory of the finite universe what this new addition to the tensor law means, the answer comes in metrical rather than physical terms: "The universe is to be regarded as a continuum which is finite (closed) with respect to its spatial dimensions." [10] The result is that his own intentions to remain consistent with the general theory are negated, a metric which should be purely relative is made to prescribe how physical objects must move and is given an absolute status again, and the way is prepared for that final complete mathematization of physics which he is later forced to repudiate. For with De Sitter,

Einstein's mere slip of practice, is accepted frankly in theory; with Weyl another frankly accepted gulf between what mathematics prescribes and physics reveals is permitted; and with Eddington mathematics is in theory set up as primary, and physical categories are relegated more and more to the rôle of mere symbols. Thus what began in Einstein's theory of the finite universe as an amendment which was to render physical theory consistent with the thoroughgoing physical demands of the general theory, ends with Eddington in a complete repudiation of those demands. Such are the unfortunate consequences which arise when man allows the powerful forces present in contemporary mathematical instruments to slip for a moment out of the controls which physical considerations prescribe.

But Einstein's failure to define finiteness in physical terms is easy to understand. Our traditional theory of matter does not provide any meaning for such finitude; it provides no physical object to prevent one part after another of matter from moving off from the rest to bring physical nature into an unstable state. Obviously, an amendment introducing a new physical entity into our traditional theory of matter is necessary. It was his failure to note this consequence,—a consequence which we have discovered that he overlooked in connection with the necessary referent for atomicity and motion,—that left him no alternative but to conceive of finitude in geometrical terms. We shall find that information given by the general theory itself, without any of these assumptions concerning stability and boundary conditions, specify what this amendment to the physical theory of nature must be. Since the general theory has been verified, we shall leave further discussion of this subject until we consider the necessary consequences of the general theory.

It is to be noted that Einstein's theory of the finite universe introduces a dissymmetry between the spatial and temporal factors of space-time. Each spatial dimension returns upon itself, whereas the temporal one does not. For this reason it is known as the theory of the cylindrical universe. De Sitter has said that this is inconsistent with the special theory since it introduces an intrinsic difference in space-time between space and time, which prevents one from regarding a temporal dimension in one frame as a spatial one in another. Certainly if the temporal dimension is infinite whereas the spatial ones return upon themselves, the one cannot be taken for the other. But this objection is fatal only if space-time has an objective status. A consistent application of the general theory removes such a possibility. There is no one space-time common to all frames which gives a different assignment of its dimensions between space and time in different frames. All space-time is relative. The space-time of one frame is as different from the space-time of another as the space or time is different. There does not seem to be any inconsistency, therefore, in a theory in which the space-time of any frame has spatial dimensions which return upon themselves and a temporal dimension which does not. But this, it is to be noted, is only the case with a theory of the finite universe which is consistent with the general theory and defines the finiteness of nature, which transcends all dependence on reference frames, in invariant physical terms. If space-time or a purely geometrical principle, is used to define finiteness, space-time becomes invariant and De Sitter's criticism holds.

2. *De Sitter's Spherical Universe*

A second theory of the finite universe was developed by De Sitter.[19] He noted that one can regard the temporal as well as each spatial dimension as returning upon itself. We can understand, therefore, why it is referred to as the theory of the spherical universe. This makes the spatial and temporal dimensions intrinsically identical. In the light of the consideration of the previous paragraph this does not seem to possess the advantage over the cylindrical universe that many have imagined.

In De Sitter's universe, space-time is regarded in theory as possessing a structure which is not completely defined by matter. It has been supposed, for this reason, that one who insists, as our analysis of the meaning of the general theory reveals that we must, that all geometry and chronogeometry is relative and is defined in physical terms, is forced to accept Einstein's rather than De Sitter's theory, regardless of what experiment may show. This does not appear to be necessary. For we have indicated that notwithstanding the direct opposition between them in theory, the difference in actual practice is only one of degree. Einstein, who correctly insists upon a thorough-going physical theory, has come no nearer to defining his finite limiting condition in physical terms than has De Sitter who frankly admits that it is not so defined. What is needed is a physical amendment and addition to our traditional theory of matter which will provide a physical meaning for finiteness, and the impossibility of any object moving away from the rest of matter. When this addition is discovered, it may be the case that the type of relative geometry which appears in any frame exhibits the properties of De Sitter's rather than Einstein's theory. In this case the relative metric of De Sitter's theory could be reconciled with the thoroughgoing physical demands of the general theory upon which Einstein so correctly insists. We shall find, however, that by proceeding from the general theory itself, without any presuppositions concerning boundary conditions and stability, it is possible to arrive at a finite theory which meets the difficulties for which Einstein's and De Sitter's theories were introduced, without beginning with any experimentally unverified assumption concerning geometry. This is possible in a theory which does not use geometrical principles to define finiteness.

The Extension of the General Theory to Electro-Magnetics

1. *Weyl's Unitary Field Theory*

The next generalization beyond the general theory was made by Weyl.[11] The need of such a generalization was evident. The general theory aimed to put our laws of nature into such a form that they hold for any frame of reference whatever. It had done this for mechanics only. It was desirable to do it for electro-dynamics as well. The tensor equation of the general theory provided a starting point. This equation contained ten variables the values of which, in a given frame, define the gravitational and metrical field of that frame. Since electro-magnetic phenomena exhibit themselves in the same regions in which gravitational and metrical fields are present, and since molar masses are known to be constituted of the electrons and protons of electro-magnetic theory, it follows that the relative gravitational, metrical, and electro-magnetic characteristics within a given frame ought to merge into a single field for that frame. In other words there must be a tensor equation which contains, in addition to the ten variables of the general theory, other variables, the values of which define the electro-magnetic character of the field of a given frame. Note the beauty and value of such an equation. Not only would our laws of mechanics and electro-dynamics apply to objective nature itself, but these phenomena would be merged and unified into a single system which might very well reveal new connections or identities between them, and throw new light upon their differences.

But is it possible to get a law which will remain invariant in form, and specify the additional number of variables? We have noted that the number of dimensions of space-time in any frame determines the number of g_{ik}'s or variables which the tensor contains. For a four dimensional space-time there are ten g_{ik}'s in a symmetrical tensor. Since no physical considerations justify more than four dimensions the possibility of gaining the additional number of variables by increasing the number of dimensions in relative space-time does not seem to be warranted.

Mathematical considerations suggested another possibility to Weyl. Einstein's general theory unifies the gravitational and metrical fields by introducing variable metrical curvature. Stated in terms of the physical principles which define a given geometry in terms of measuring rods and their behavior, this means that parallelism holds only over infinitely short displacements of the rod. Hence parallelism for finite distances is never assured *a priori* by the nature of space, but is a character of relative space which can only be determined empirically by an observation of the actual way in which measuring rods behave when they are moved around in the region of a given frame. It was assumed, however, in Einstein's general theory that the length of the rod remained the same for any displacement whether finite, or infinitely short, in extent. This is expressed in technical terms by the statement that there is metrical, but not gauge curvature.

The suggestion is quite obvious. Can we not gain the additional number of variables by introducing gauge as well as metrical curvature?

Weyl followed this line of thought. Note that it has the advantage, from the mathematical point of view, of placing all the *a priori* principles of geometry upon an infinitesimal basis. No equalities either of direction or length are assured when a rod is moved a finite distance. Thus Weyl writes in his original paper that "the fundamental conception on which the development of Riemann's geometry must be based . . . is that of the infinitesimal parallel displacement of a vector." [11] No geometrical identities of length or direction can be counted on for more than infinitely short regions. Stated in more concrete terms this means that a rod which is carried out around a large circle and back to its starting point will not have the same length at the end of the journey as at the beginning. The thesis that geometrical principles which hold for infinitely short displacements, do not hold for longer displacements is expressed in technical terms by the principle that the differential equations describing the infinitesimal properties of space-time are not integrable. Thus Weyl writes that "It is only in Euclidean 'gravitationless' geometry that integrability obtains." [11]

The remarkable thing is that when Weyl developed the

tensor which would remain invariant upon such a basis, he found himself with a tensor containing not merely the ten variables of the general theory but with four others termed K_i's as well. Moreover, when he identified these K_i's with the electro-magnetic potentials he found that the identification was valid since the extra equations to which his deductions gave rise turned out to be none other than Maxwell's equations for the electro-magnetic field. Thus science found itself possessing a unitary field theory; and the greatest advance since the discovery of the general theory was at hand.[18]

2. *Eddington's Mathematical Theory*

At this point Eddington made another suggestion.[19] He noted that the gauge curvature which Weyl introduced is not the most general that is possible. A rod may vary in its length when rotated about a given one of its points, as well as when transported over a finite distance. With this generalization a more complicated tensor results requiring many more variables to be specified before the geometry of the space-time of a given frame is determined. Concrete values for all these variables do not appear in our observable world.[5] Hence, Eddington's theory lacks physical significance.

In fact, we may well ask the question whether Eddington's theory must not be rejected since it gives rise to deductive consequences, for which we find nothing to exist? But Eddington did not reason in this fashion. In order to save this very brilliant mathematical generalization and explain the difference between the prescriptions of theory and the deliverances of physical fact, he added another hypothesis to the effect that the mathematical system is the real world and the physical world is the phenomenal one. But it was still necessary to explain why there is such a difference between the metric of one and the metric of the other. But this did not cause Eddington to falter. Another hypothesis was introduced. Reality is not merely a system of relations with electrons as mere symbols, but there are souls which select out of the possibilities which the most generalized mathematics permits, the particular restricted part which is the metric of our actual world.[21] In this fashion a theory which necessitates that only physical objects and their motion are objective and that all mathematical structure is relative was made the basis of the thesis that electrons are symbols, and only relations are objective. Thus relativity physics was made the basis of an argument to mysticism.[21]

But if it is necessary to add on so many hypotheses in order to adjust one's theory to fact, is there not some doubt concerning whether the theory is worth saving? Why introduce so many new theories to explain a given theory when there is nothing in observed nature which the theory itself explains? And especially when the result is the very antithesis of the meaning of the verified general theory which is the certain ground from which all these flights of the mathematical and speculative imagination have proceeded.

But it is not necessary to go outside of Eddington's theory to discover that his thesis of the primacy of relations and psychological categories over physical one's is not tenable. For it seems to be impossible to define the axioms of his own most general geometry in terms of the principles of the philosophy which he lays down. An examination of these axioms, which appear at the end of his book, "The Mathematical Theory of Relativity,"[19] will suffice to make this clear. They refer to "displacements" "carried by parallel displacement" from point to point. Later he defines a displacement as "the comparability of proximate relations." The latter definition does not seem to mean very much, but even if we admit that it has meaning, it certainly cannot be the one which is necessary to satisfy the axioms of his own absolute geometry, for relations whether similar or dissimilar, proximate or non-proximate, cannot be carried; only physical objects, or in this instance, measuring rods, possess this property. It becomes evident, therefore, that physical objects and physical motion is really primary even in Eddington's extremely mathematical theory. Moreover Eddington accepts Einstein's definition of simultaneity for spatially separated events. This, as we noted in conjunction with the special theory, is fatal to his mathematical and psychological philosophy since it presupposes absolute propagation and the physical theory of nature. Of all the students of relativity theory, Whitehead is the only one who can reject a physical interpretation of it, since he is the only one who rejects Einstein's definition of simultaneity and attempts to find a meaning for the relativity of space and time in non-physical terms.[22] The result, however, is a theory of relativity entirely different from Einstein's, or any of the generalizations beyond Einstein.

It happens to be the case, however, notwithstanding its derivation of Maxwell's equations, that Weyl's theory also reveals a discrepancy between the geometry of theory and the geometry of fact.[35] At first Weyl regarded his theory as possessing the same complete and perfect physical validity as Einstein's general theory. Further considerations revealed, however, that this is not so. It gives rise to deductions which are not entirely in accord with fact. For example, frequencies of identical vibrating atoms should vary with their past paths; experiment does not confirm this conclusion. Thus Weyl was forced to draw a distinction between his theoretical gauge and the natural gauge of Einstein's general theory.[16] Evidently, even in Weyl's case pure mathematics is guiding man forward to theories a little faster than fact will permit.

3. *Einstein's Unitary Field Theory*

It becomes evident, therefore, that we must return from the theories of Weyl and Eddington to the secure foundations attained in the special and general theories. Weyl's generalization remains a genuine achievement and possesses physical significance in part at least since it gives rise to Maxwell's equations. But if the general theory is true we must look to a different formulation for the unity of gravitation and electricity. For the general theory indicates that all geometrical principles are relative and must be defined in physical terms. If this is true there can be no distinction, such as Weyl and Eddington draw, between the gauge of theory and the gauge of nature. No mathematical formulation is permissible which physical nature does not define.

Evidently Einstein has sensed this point. For in 1926 he found himself forced to repudiate the course which Weyl and Eddington followed,[13] and in December of 1928 he brought out a new unitary field theory which gains the additional potentials within the geometry of the general theory. This is accomplished by rejecting the thoroughgoing infinitesimal geometry of Weyl and Eddington, retaining the gauge constancy of the general theory, and introducing a meaning for parallel displacement over finite distances, by making the differential equation which defines parallelism integrable.[14] Thus Einstein writes "The new unitary field theory is based upon the following mathematical discovery: There are continua with a Riemannian metric and distant parallelism which nevertheless are not Euclidean."

To appreciate the uniqueness and full significance of this theory it is necessary to understand certain problems not referred to by Einstein, to which the general theory gave rise. This brings us to a consideration of certain necessary consequences of the special and general theory which were discovered before Einstein's unitary field theory was known.

NECESSARY CONSEQUENCES OF THE SPECIAL AND GENERAL THEORIES

The most interesting and important question now arises. What change in our conception of the nature of things does the theory of relativity necessitate? In other words, what else must be true because the special and general theories are true?

Our previous analysis of these theories provides us with part of the answer. The physical theory of nature in a kinetic atomic form must be true. For after introducing and recognizing more relativity than even the most imaginative speculative mind has ever conceived, Einstein's discoveries reveal that there is something absolute in nature, remaining objective and invariant through all the relativity which exists, and that this absolute factor is matter and motion.

Let us recapitulate the reasons for this conclusion. First, the theory of relativity grew out of and presupposes the electro-dynamic theory of Lorentz, and this is a kinetic atomic or electron theory.[2] Second, the most elementary and essential principle in relativity physics is the principle of the relativity of simultaneity, and this principle is defined in terms of a physical motion or propagation which must be taken as undefined, and hence as absolute. Third, all the relativity of Einstein's theory is defined in terms of rods and clocks and reference frames, all of which are physical objects. Even events in his theory are conceived in physical terms. Fourth, the general theory takes the subject-matter of mechanics, which is matter and motion and succeeds in stating it in absolute terms. Finally, in doing this it reveals not only space and time, but chronogeometry or space-time to be relative, and that even their relative character "is defined at every point by the matter at that point and the state of that matter." Incidentally, notwithstanding its apparent philosophical incompatibility with the general theory, even the extremely mathematical and speculative theory of Eddington does not escape from the primacy of matter and motion in the definition of its own axioms. The conclusion is inevitable: nature is physical and contains motion. Furthermore, motion is presupposed in defining space-time. Since only motion which is defined in terms of space-time is relative, it follows that the theory of relativity not only provides a meaning for, but necessitates the existence of absolute motion. The relative motion which is measured in the given relative space-time of a reference frame is not the only type of motion which exists.

Motion and Metrical Uniformity

From the fact that nature is physical and contains change which is motion,[83] it follows necessarily that nature is kinetic atomic in character and that a referent in addition to the moving microscopic particles exists. Parmenides proved this for all time, as our analysis of Greek science has indicated. In this connection it is of philosophical interest to note that E. Meyerson quotes Einstein as saying that it is "a shallow view of science which leads it to reject the existence of atoms." [84]

The importance of Parmenides' proof, like the beauty of his analysis, is such, that it bears repeating. If there is no referent other than the stuff which moves, change due to motion is impossible for two reasons: there cannot be motion, and there cannot be atoms. For motion involves the transition of stuff from where it is to where it is not. If nothing but the stuff which moves exists, there can be no where-it-is-not, and hence motion is impossible. Nor can this difficulty be met by regarding matter as many rather than one, for two reasons. First, the motion of many particles involves a where-it-is-not as much as the motion of one. A difficulty is not met by multiplying it many times. Second, there cannot be many particles. For manyness requires some referent to go between the parts, and distinguish one part from another. If nothing but the stuff to be divided exists, no such additional factor is present and hence atomism is impossible.

It is to be noted that the last argument eliminates the possibility of defining atomic motion in terms of the relation of the microscopic atoms to each other. Of course relativity physics itself condemns such a theory, for we have indicated that it involves absolute motion. But even if this were not immediately evident, a referent other than the moving microscopic particles must exist. For even a relative motion involves many substances; and their manyness, wholly apart from their motion, is meaningless and impossible, unless there is a referent other than the substances themselves.[80]

The point in this argument of Parmenides is not so much that an intervening medium is necessary, as that a basis for distinguishing between one atom and another is required. In a theory which regards the different substances as made of the same kind of material, the category of stuff gives only the respect in which the atoms are identical or one; it cannot prescribe the respect in which they are different or many. Stated positively this means that the individuality of one atom can be distinguished from that of any other, only in terms of its unique relation to some common referent. If nothing but the microscopic particles themselves, is supposed to exist, no such referent is present, and atomism is impossible. Since relativity physics reveals that matter in motion exists and is primary, it follows that a referent in addition to the moving microscopic particles must also exist.

Once this is recognized it becomes evident that Einstein's discoveries necessitate an amendment to the traditional physical theory of nature, to provide a new referent for atomic motion. For with the Michelson-Morley experiment and the acceptance of the special theory, the traditional referent introduced by the Greeks and accepted by Galilei and Newton was revealed to be non-existent. With the general theory, not merely space-time, but any possible geometrical or chrono-geometrical entity, was revealed to be out of the question. Not only was space-time revealed to be relative, but even its purely relative character is conditioned by matter and motion. Certainly that which is defined in terms of matter and motion cannot be used to define what we mean by matter and motion. Moreover the inadequacy of any possible mathematical structure indicates that this new referent must be a physical entity.

This should not surprise us. In fact, the difficulty is to explain why we have not discovered this before. The impossibility of defining the motion of matter in terms of a relation to space was pointed out in ancient times by Zeno,[30] and reëmphasized in the modern world by Berkeley and Mach. The slightest reflection should indicate that such a theory involves a vicious muddling of physical and mathematical categories. How a disembodied conceptual form can determine the direction of motion of a physical particle is hard to understand. One might as well try to derail the Twentieth Century Limited by conceiving of the equation of a disembodied open switch in its pathway. Only the worship of a dogma can explain the existence of such nonsense in science. Yet how often we hear scientists repeat this dogma with bated breath: Nothing is truly scientific which is not mathematical. This thesis is not valid unless the mathematical theory of nature is true, and there has been no justification for this conclusion since Thomas Aquinas rejected Platonism in the thirteenth century. All the mediaevalism is not in theology.

The question now arises concerning the nature of this new referent. Additional necessary consequences of the general theory provide an answer to this question. A single sentence suffices to express them: Nature exhibits a metric in every frame, from which we have measured astronomical distances, which possesses approximate macroscopic metrical uniformity and constancy; and this metric is conditioned by matter. The latter part of this statement has been demonstrated in conjunction with our analysis of the general theory. The former follows because the general theory cannot be true unless the principles of measuring are valid; and they necessitate the existence of such a metric.

The consideration of what is involved in the measurement [24] of astronomical distances will suffice to make this clear.[29] Consider a distance here on the earth. The measurement of its length presents no problem for one who accepts the physical theory of nature. One merely takes a conventionally chosen physical object called a measuring rod, and notes how many times it must be laid down end to end to cover the distance in question. Incidentally, it is only if one accepts the kinetic atomic physical theory that there is any meaning for the distinction between a rigid and a non-rigid rod. Otherwise, even a partial answer cannot be given to the question concerning why our laws do not hold as well for measurements made with an angle worm, as with the coddled instrument in the Bureau of Standards. It is one of the many merits of Whitehead to have made us reflect on these questions;[24] undoubtedly much more reflection is necessary. There is too much of a tendency to slide over such difficulties with facile statements concerning what happens in practice, as if the rationale of practice itself does not need to be made explicit. I am very much indebted to my colleague, Dr. Hoskyn, for increased appreciation of this point.

Now turn from a terrestrial distance to an astronomical one. The determination of its length is not so easy. One cannot carry one's rod to the sun and then go cruising out through the intervening region to Sirius, moving one's measuring rod along as one goes. The difficulty is met by establishing a relation of equality or congruence between the directly measurable local distance and the inaccessible celestial one. To establish this relationship it is necessary to appeal to geometrical principles. Moreover, if the results are to be valid these principles must apply to the intervening region which relates the two distances in question.

If the metrical structure of this region were not at least approximately uniform and constant two difficulties would occur. First, the geometrical principles to which we appeal at one time would not be those which we use at another. Were this true, the values of an observation made in one century should not continue with those of another to make sense. In short, the values in question would be incommensurable. Second, even in the single instance we should not know what geometrical rules to use until we had determined the metrical structure of the intervening region at the time. But one cannot determine this structure without making astronomical measurements. Thus if the metric of space is not approximately constant one finds one's self in the peculiar predicament of not being able to make a single astronomical measurement until one has made a large number of such measurements. The experience of astronomers does not confirm this consequence. It must be assumed, therefore, that the premises which necessitate it are false and that at least an approximate metrical uniformity and constancy extends over macroscopic regions of the space of any frame used for astronomical measurements.[29]

In this connection Whitehead pointed out that it is difficult to understand how the general theory can be true.[28] For according to the latter theory the metric of space is conditioned by matter. To quote Einstein's own words, "According to the general theory of relativity the metrical character (curvature) of the four dimensional space-time continuum is defined at every point by the matter at that point and the state of that matter. Therefore, on account of the lack of uniformity in the distribution of matter, the metrical structure of this continuum must necessarily be very complicated." [10] Hence Whitehead argued, since the relations between material objects are contingent, according to the traditional theory of matter, it follows that metrical variability should exist, and the measurement of astronomical distances as actually carried on should be impossible. He writes, "I cannot understand what meaning can be assigned to the distance of the sun from Sirius if the very nature of space depends upon casual intervening objects which we know nothing about. Unless we start with some knowledge of a systematically related structure of space-time we are dependent upon the contingent relations of bodies which we have not examined and cannot prejudge." [28]

In order to meet this difficulty and to explain why we have not discovered metrical variability before, Einstein stated the general theory in terms of finite distances, and conceived of the metrical variability as a microscopic exception to the general macroscopic uniformity. This seems to be the case. Certainly astronomical measurement would be impossible in its present form were the macroscopic uniformity absent, and the general theory cannot be true unless variability exists.

Many have supposed that this leaves everything in a satisfactory state. Quite the contrary is the case. The difficulty on which Whitehead placed his finger, has been merely shifted into another form in which it becomes all the more acute. Instead of asking why microscopic variability is so rare, we now have to inquire why macroscopic uniformity is so prevalent. In other words, the rule rather than the exception to it presents the difficulty. For if the metric of space is conditioned by matter as the general theory necessitates, it follows from our current theory of matter that the general macroscopic metrical uniformity and constancy should not exist. In short, exactly the opposite of what Einstein suggests and the facts indicate, should occur.

This becomes clear the moment one remembers that the theory of relativity necessitates a relational theory of space which is defined in terms of matter, which in turn means that the metric of space must be defined in terms of relations between physical objects; ultimately, since we have demonstrated that relativity physics necessitates a kinetic atomic theory, this means, in terms of relations between moving atoms. In such a theory metrical variability exists when the relations between the atoms change, and metrical uniformity when they do not. The variation of physical potentials on the molar level to which the tensor laws refer, is but an exhibition of this basic atomic motion expressing itself in compound bodies and their molar motion and interrelations.

Since the atoms of the universe of the relativity theory are kinetic in character the relations between them are continuously changing. Hence metrical variability should be the general rule. Metrical uniformity should be the microscopic exception, holding only for infinitely short increments of atomic motion. Hence if the traditional theory of matter is retained and the metric of space is conditioned by matter, it follows that the macroscopic metrical uniformity and constancy necessary for measuring cannot exist.

The Functional Theory of Whitehead

The first serious attempt at a solution of this difficulty was made by Whitehead.[22] He concluded that the physical theory of nature must be rejected and that science must be reared upon a theory of first principles which makes the principle of becoming fundamental, and admits a space-time relatedness which is not conditioned by matter. This modern edition of the functional theory of nature conceives of nature as a vast extensive process. It is by abstractions from this monistic process that our scientific concepts are derived. The first and least falsifying abstraction consists in the assignment of different parts of the "passage of nature" to different classes by means of the relation of simultaneity. The relativity of time arises from the fact that observed nature is too complex and ambiguous to insure that this assignment must always be made in the same way.[30]

In each time system the complex of events exhibits two main factors: first, a certain constant uniform structure between its parts, termed space; and second, many adjectival permanences termed "sense objects", which are the content that appears within this structure.

Since there is a meaning for space only in a given time-system, a basis is provided for the relativity of space; and since the structure of space in a given time-system is constant and uniform and independent of the sense data and their "controls" which constitute molar objects, the uniformity and constancy necessary for measuring exists, notwithstanding the changing relations of objects. Thus, his theory provides a referent for molar motion in a given time-system (but not, it is to be noted, for kinetic atomic motion which is independent of any time-system [28]) and a basis for measurement, while also enabling us to accept the relativity of space and time which the Michelson-Morley experiment necessitates. This is the unique achievement of his philosophy with reference to the findings and difficulties of current inorganic science.

The originality of Whitehead's conception of relativity has not been fully sensed. His doctrine is not a philosophical interpretation of the theory of Einstein which we have traced in this chapter and which physics accepts, but a thoroughgoing independent and original theory. Not even Einstein's definition of simultaneity, which is the most elementary and essential concept of the special theory, is accepted by Whitehead. The former admits the existence of an immediately given simultaneity only for spatially coexistent physically defined events; and uses a physical motion or propagation to define the simultaneity and relativity of spatially separated events. This is but another way of saying that the theory of Einstein, which Weyl, De Sitter, and Eddington accept, is a physical theory. Only if physical categories are primary can Einstein's theory be justified. Whitehead on the other hand maintains that there is an immediately given fact of simultaneity not merely for spatially coexistent events, but for the whole of "discerned and discernible" nature.[22] This follows, of course, only if one defines the categories of physics in terms of those of a phenomonalistic philosophy of events, and can convince one's self that a meaning for "now" for the whole of nature is intuitively given. Relativity arises for Whitehead not from the fact that a definition of the simultaneity of spatially separated events in terms of light propagation necessitates that those events which are simultaneous for one physical frame cannot be simultaneous for another; but in the circumstance that the passage of nature is too ambiguous to insure that the intuitively given simultaneity for the whole of nature, is the same for one person as for another. It is because of this that Whitehead gets a meaning for relativity which does not depend upon physical reference frames and light rays, and is the only student of contemporary physics who is entitled by his own premises to reject the physical theory of nature.

The reader must decide whether a doctrine which places the source of relativity in an intrinsic ambiguity in nature's passage and which admits all the relativity which psychological immediacy entails can provide a foundation for scientific findings. Certainly Einstein would maintain that it can not, on the ground that it does not enable one to decide in the concrete instance whether one has relativity or not, or what the precise amount is.[6] We must also ask ourselves whether a theory which begins with the simultaneity of heard noises and observed colors is not driven in the end, in order to get order out of such a phenomenal and chaotically relativistic world, to define the comings and goings of these immediate events in terms of the physically defined events and relativity with which physics deals and with which Einstein is concerned. The writer can find little in the history of modern science or in current scientific developments or philosophical analysis to indicate that this is not the case. If so the physical theory of nature must be primary. Even so the importance of Whitehead's work remains for he is the first one to see the real difficulties concerning atomicity, motion, and measuring to which Einstein's theory of relativity has given rise, and the only one who has paid what it costs to reject the physical theory of nature in order to meet them. Moreover, when the physical theory of nature is carried to its necessary consequences we shall find ourselves in agreement with Whitehead on the major general outlines of philosophy. The differences at this point are not as far-reaching as they first appear.

Notwithstanding the solution of the problem of measuring, which it provides, Whitehead's theory comes at a terrific cost. It is necessary for him to reject the general theory. This follows because the independence of matter from space-time which his theory entails, necessitates the rejection of the doctrine that the metric of space is conditioned by, and varies with, the distribution of matter. To reject this is to deny Einstein's law of motion and his tensor equation for gravitation. The reason for this decision by Whitehead can be appreciated. He preferred an adequate theory of measuring to a law for gravitation which made measurement impossible. If a choice is necessary at this point it must be made in his favor, for a science of mechanics without the general theory is possible, but mechanics without measurement is nonsense.[28]

But the experimental findings do not permit us to make such a choice. Evidence for the principles of measuring and the general theory exists. Information from astronomy indicates not merely that the general theory is valid but that the specific doctrine of a variable and heterogeneous metric must be accepted. What the facts call for is a theory which can prescribe the macroscopic metrical uniformity necessary for measuring and also admit the thesis that space is conditioned by matter.

The conclusion is inescapable. Since the verified special and general theories reveal that the metric of space is conditioned by matter, and a macroscopic metrical uniformity exists which matter, as traditionally conceived, cannot condition, it follows that our traditional theory of the capacity of matter to produce structure and constant order in nature is false, and must be amended to meet new evidence. It is only if this possibility is overlooked that Whitehead's rejection of the physical theory of nature appears to be a necessary consequence of the difficulty concerning measurement which he has revealed.

The question arises immediately concerning what this amendment must be. An examination of the metric of any reference frame in this universe, with special reference to the physical entities which are necessary to condition it, should provide an answer to this question. The requirements for measuring and the general theory indicate that it possesses the two characteristics which Einstein has suggested. An approximate general uniformity and constancy extending over macroscopic distances is interspersed with local microscopic heterogeneity and variability.

The basis for the microscopic metrical variability is already at hand in the traditional atomic theory. For, if we mean, as we must in a physical and relativity theory, that space is a relation between the atoms and that metrical variability is a change in this relatedness, it follows from the kinetic properties of the microscopic particles that metrical variability must exist. It follows also that the material basis for the constant macroscopic metrical uniformity can not be found in the microscopic atoms. For, if their properties are such as to necessitate a variation in their interrelations, then relational constancy must have some other source.

We discover, therefore, that something other than the traditional microscopic atomic entities must exist in this universe. In fact, we have but to note what is required to impose a constant macroscopic metrical uniformity upon the local variable relatedness of the microscopic atoms to discover what this additional factor is.[28]

First, it must be physical. Otherwise it would be necessary to reject the general theory and its doctrine that space is completely conditioned by matter. Also, if it is to cause the microscopic atoms to compensate their variable relatedness so that a constant uniformity extending over great distances in nature is preserved, it must change the direction of their motion. This calls for the presence of an external force which only a physical object can provide.

Second, this physical entity must congest and surround all the microscopic atoms of the whole of nature. Otherwise, they would be merely crowded out into some other referent for their motion, and macroscopic metrical variability would be the rule.

Third, this physical object must be an atom rather than a compound substance. Otherwise some referent other than it would be required to provide a meaning for the distinction between one of its parts and another, and the old difficulty over atomicity would recur.

We have but to bring these different requirements together to discover that this universe must be constituted not only of the moving microscopic atoms of the traditional atomic theory but also of one large physical macroscopic atom, spherical in shape and hollow in its interior except for its inner field, which surrounds and congests them.[28]

The Macroscopic Atomic Theory [27]

In this fashion the information which the general theory gives concerning the physical basis and metrical character of space reveals the specific character of the new physical referent which the kinetic atomic basis of relativity physics necessitates. We shall henceforth refer to this new physical theory of nature as the macroscopic atomic theory.[27]

Note how this new theory of the first principles of science meets the difficulties which relativity physics has raised. Since the shape of the macroscopic atom is different from that of the microscopic atoms, a meaning for the difference between it and them is to be had, without recourse to an additional common factor. Thus a referent is provided for the atomicity and motion of the microscopic particles without involving one's self in a circular argument. This makes it possible to define the metric of space-time in terms of relations between physical objects, and to reject all geometrical entities from the background of natural occurrences, without unsolving the problem of atomicity and motion.

Furthermore, since the macroscopic atom introduces a fixed spherical physical form which surrounds and congests the miscroscopic atoms sufficiently to impose an approximately constant macroscopic uniformity, but not so completely that their motion and the resultant variable relatedness is prevented, the peculiar combination of macroscopic metrical uniformity and constancy which measuring requires, and of microscopic metrical heterogeneity and variability which the general theory necessitates, is made intelligible in strictly physical terms. Moreover, our theory places these attributes precisely where empirical science finds them. Variable and contingent relatedness has its basis in the changing relations of the moving microscopic particles, and hence is down in the regions of the microscopically small; and uniformity and constancy is imposed from without by the changeless physical form of the macroscopic atom, and hence appears over molar and macroscopic regions in the stellar realms of astronomical space. Thus, the unexpectedly static character of the fixed stars is made intelligible, the approximate validity of the special theory is explained, and the physical demands of the general theory are reconciled with its metrical consequences.

Moreover, since the diameter of the macroscopic atom is finite and since its atomic or indivisible character prevents any of the microscopic particles from moving away from the rest never to return, it follows that the macroscopic atomic theory, which we have deduced from the general theory alone, meets the difficulties that give rise to the more speculative theories of the finite universe, without any new assumptions, and without committing the fallacy of defining motion in terms of the returning character of the spatial dimensions of space-time.

Finally, our theory entails a geometrical formulation beyond that of the general theory, and different from that of Weyl and Eddington.[29] In the first place, an adequate law must specify variables for the electro-magnetic as well as the gravitational potentials. Otherwise the compound field to which the tensor equations refer could not be made up of the fields of the microscopic atoms of electro-magnetic theory, as well as the inner field of the macroscopic atom. Second, it must give rise to relative systems of geometry or chrono-geometry which are Riemannian. Otherwise the variable potential distribution in local regions, which the motion of the microscopic atoms necessitates, could not exist. And third, it must provide a meaning for metrical uniformity or constancy over finite distances. Otherwise the macroscopic uniformity which the macroscopic atom necessitates would be equally non-existent.

It is hardly necessary to add that this is precisely the type of geometry which Einstein has discovered in connection with his recent unification of the electro-magnetic and gravitational fields.[14] He writes: "The unitary field theory is based upon the following mathematical discovery: There are continua with a Riemannian metric and distant parallelism which are nevertheless not Euclidean." In other words a geometry has been explicitly formulated which has the peculiar combination of local variability and macroscopic uniformity and constancy in its metric, which the macroscopic atomic theory introduced in order to reconcile Einstein's doctrine of the dependence of space-time upon matter, with Whitehead's doctrine of the necessity of metrical uniformity for measuring. Moreover, Einstein is giving us more and more reason for believing that this is the actual relative geometry of the relative physical fields of our universe.[29]

Note how the physical demands of the general theory are being met. Einstein's new field theory not only brings the gravitational and electro-magnetic fields together but it accomplishes this result while also providing a meaning for macroscopic metrical uniformity, within the limits of the natural Riemannian gauge of the general theory.

Since the macroscopic atomic theory was discovered and formulated independently of and previous to Einstein's latest achievement, it may be maintained that his unitary field theory constitutes additional evidence for its validity. In fact, it is difficult to understand how the metrical uniformity involved in distant parallelism can exist in a universe in which the metric of space-time is conditioned by matter, unless the macroscopic atom exists to offset the contingent relatedness to which the motion of the microscopic particles gives rise.[29]

At last we find the theory of relativity established upon secure, consistent, and sufficient foundations, in terms of the macroscopic atomic theory. The consequence of the theory of relativity with reference to our conception of the nature of things is simple when once attained: The Greeks made a mistake which the moderns have perpetuated when they introduced a mathematical rather than a physical entity to provide a meaning and basis for atomicity and motion.

Such is the lesson which nature has been trying to teach us during all our study and analysis from Zeno to Einstein. Natural processes do not take place in a geometrical entity termed space, or in a static independent continuous substance termed the ether, or in a chrono-geometrical continuum termed space-time. Instead they are the result of the motion and compounding of microscopic atoms with their fields in the macroscopic atom with its inner field. Motion is not to be defined in terms of space and time, but space and time are relative abstractions from or relations between moving objects. Matter does not move in space and time, but microscopic particles which obey the principle of being move in the macroscopic atom which obeys the principle of being. Not only is motion more primary than space and time, but the notion of eternity, as defined by the principle of being, is more fundamental than the notion of temporality.[31] When the full implications of this fact are grasped the whole complexion of contemporary thought and civilization will change.

References and Bibliography

1. A. V. Vasiliev. Space, Time, Motion. P. 47. Knopf.
2. H. A. Lorentz. The Theory of Electrons. Stechert & Co.
3. M. Born. Einstein's Theory of Relativity. Dutton.
4. J. H. Thirring. Ideas of Einstein's Theory. Methuen.
5. A. D'Abro. Evolution of Scientific Thought from Newton to Einstein. Boni & Liveright.
6. A. Einstein. The Theory of Relativity. Holt.
7. Einstein and Others. The Principle of Relativity. Methuen. Includes Eng. Trans. of foll:
8. A. Einstein. On the Electro-dynamics of Moving Bodies. Annal. d. Physik. 17, 1905.
9. A. Einstein. The Foundations of the General Theory. Annal. d. Physik. 49, 1916.

10. A. Einstein. Cosmological Considerations on the General Theory. Sitz. d. Pr. Ak. d. Wiss. '17.
11. H. Weyl. Gravitation and Electricity. Ibid. 1918.
12. A. Einstein. Sidelights on the Theory of Relativity. P. 28ff. Dutton.
13. A. Einstein. Math. Annal. 97. P. 99.
14. A. Einstein. Zur Einheitlichen Feld Theorie. Pr. Akad. d. Wiss. 1929 I, Also 1928 XVII.
15. H. Weyl. Space-Time-Matter. Methuen.
16. H. Weyl. Raum Zeit Materie. Fünfte Auflage. Springer. 1923.
17. H. Weyl. Was ist Materie? Springer. 1924.
18. H. Weyl. Philosophie der Mathematik u. Naturwissenschaften. Oldenbourg. 1927.
19. A. S. Eddington. The Mathematical Theory of Relativity. 2d Ed. Cambridge Press.
20. A. S. Eddington. Space, Time and Gravitation. P. 197. Cambridge Press.
21. A. S. Eddington. The Nature of the Physical World. Macmillan.
22. A. N. Whitehead. The Concept of Nature. Cambridge Press.
23. A. N. Whitehead. The Principle of Relativity. P. 58ff. Cambridge Press.
24. A. N. Whitehead. Process and Reality. 491—. Macmillan.
25. B. Russell. The Principles of Mathematics. Cambridge Press.
26. B. Russell. Introduction to Mathematical Philosophy. George Allen & Unwin.
27. F. S. C. Northrop. The Macroscopic Atomic Theory. Journ. of Phil. XXV No. 17.
28. F. S. C. Northrop. Two Contradictions in Current Physical Theory. Proc. Nat'l. Ac. of Sc. Ja. 1930.
29. F. S. C. Northrop. Einstein's Unitary Field Theory and the Macroscopic Atomic Theory. Monist. 1930.
30. F. S. C. Northrop. Philosophical Consequences of The Theory of Relativity. J. of Phil. XXVII.
31. F. S. C. Northrop. The Relation between Time and Eternity, etc. Proc. 7th Int. Cong. of Phil. Oxford. 1930.
32. F. P. Hoskyn. The Problem of Motion. Jour. of Phil. XXVI.
33. F. P. Hoskyn. The Adjectival Theory of Matter. Jour. of Phil. XXVII.
34. E. Meyerson. La Déduction Relativiste. P. 62. Payot.
35. H. Reichenbach. Philosophie der Raum-Zeit-Lehre. W. de Gunther.

CHAPTER III

QUANTUM AND WAVE MECHANICS AND THERMO-DYNAMICS

The history of optical and atomic physics has been the story of a shift back and forth between a continuous and a discontinuous theory of natural phenomena. In the seventeenth century Newton and Huygens presented evidence in support of both the emission or discontinuous, and the wave or undulatory theories of light. The triumph of the physical mechanics of Newton over the mathematical philosophy of Descartes placed the discontinuous theory in the ascendency and threw the weight of the mathematical physicists in favor of such a theory of optics and electricity. Hence, at the opening of the nineteenth century,[1] when Young and Fresnel gave evidence in favor of the wave theory of light, and Faraday gave similar support to a field theory of electricity and magetism, they were opposed by the authorities of their time. The notion of action at a distance, and thoroughgoing discontinuous conceptions ruled the field.

Into this situation came Clerk Maxwell. After stating Faraday's ideas in mathematical form, he found that the electro-magnetic ether which Faraday demanded, also possessed the properties of the optical ether which Young and Fresnel had established. Thus optics and electricity and magnetism were unified, and science was as certain of the adequacy of a continuous theory as it had been sure of the discontinuous theory a few decades before. The pendulum had swung to the other extreme.

Meantime chemistry and thermo-dynamics had been driven to the chemical atomic theory of matter, and the kinetic atomic theory of gases and heat. It was necessary to face the question concerning the relation between these discontinuous theories and the continuous theory of electro-magnetics. Kelvin and Maxwell attempted to answer this question in favor of the latter type of theory, by defining the atom in terms of a vortex or pucker in the continuous ether.

But the field theory had no more than made itself supreme by this bit of manipulation before Lorentz was faced with the task of generalizing Maxwell's equations so that they applied to bodies in motion. The result was the discovery of the atomicity of electricity. To anyone who knows his Greek natural philosophy, this comes as no surprise. One would expect a continuous physical theory to work as long as one did not face the question of change or motion, and to break down the moment one did. For the Greeks proved that nature cannot be physical and contain change unless it is kinetic atomic in character. Electro-magnetics learned this lesson with Lorentz, as chemistry had learned it earlier with Lavoisier and Dalton.

Soon after this Sir J. J. Thomson isolated the electron experimentally, and Millikan determined its electrical charge. Man gave his time to a study of the manner in which electrons and protons combine to form chemical atoms. The triumph of the discontinuous theory was supposed to be complete. Moreover the Brownian movement gave almost direct observable evidence of the existence of atoms.[2] This victory was accentuated when the Michelson-Morley experiment and the theory of relativity forced electro-magnetics to reject the ether. The last vestige of continuity was thereby removed.

At this stage, relativity theory absorbed the attention of men. Its mathematical emphasis and chrono-geometrical character caused uncritical scientific minds to suppose that the continuous theory was again established. Nature is a four-dimensional continuum in which an object is but a series of static event particles taken as at rest or in motion according to one's frame of reference, it was said. In this instance, however, the appearance of the primacy of the continuous was but an illusion, an illusion to be sure from which many of our scientists have not yet recovered, but nevertheless an illusion. For careful analysis reveals that the theory of relativity is not talking about an absolute space-time continuum at all, but about many physical objects and their absolute motion. It is with a pluralistic or discontinuous physical universe, which must be kinetic atomic in character that the theory of relativity is concerned.

QUANTUM THEORY

New developments in thermo-dynamics and in atomic theory, as it bears on the findings of spectroscopy, were to emphasize this consequence to an unexpected and astonishing degree.[3] The first hint of the need for a new theory came when Wien deduced the law for black body radiation which must hold if traditional statistical and thermo-dynamical principles are true, and the experiments of Lummer and Pringsheim revealed that Wien's law does not hold except for the special case of high frequencies. It was necessary, therefore, to modify the statistical basis of thermo-dynamical theory. The German physicist Max Planck attacked this problem. He noted that the traditional expression for the mean energy is not unique. So he altered this expression by assuming that the energy is not distributed continuously, but is divided into a discrete number of finite elements which are distributed at random among a finite number of possible positions. From these presuppositions he deduced a law for black body radiation which fits the facts, and discovered a certain constant which is designated by the symbol h.

Note what had happened. Energy as well as electricity and matter is discovered to be atomic in character. It cannot be given out in any amount from zero to infinity, but appears only in specific finite units of equal amount. Furthermore, the statistical theory, in terms of which the necessary consequences of random distribution are defined, is stated in terms of finite rather than infinitesimal units. Thus, it is to be noted that quantum theory is fundamentally discontinuous in two respects. Not only energy itself but the background in which it is distributed is conceived in finite and discrete terms. Since none of the more recent developments escape from these elementary ideas, it follows that current atomic theory is discontinuous in character.

So much for the radiation of black bodies in its bearing upon our conception of energy. But it was noted that theory can not stop at this point. If other verified laws are to hold, the atom or element of energy ϵ, must be set equal to the product of the frequency and a certain constant *h,* known as Planck's constant, which Plank's new statistical theory had revealed.[3] This constant, *h,* is known as the quantum of action, since, as the equation indicates, it is determined in terms of energy and time.

The matter rested here until 1905, when Einstein pointed out that these conceptions entail a quantum theory of radiation. In other words, radiation possesses a quantum-like structure when it is propagated through a medium. With this consequence, not merely the ether, but the continuous character of the traditional wave propagation was rejected. Thus another aspect of the continuous theory of nature went into the discard.

Meanwhile the physical chemists were developing their theory of the chemical atom in terms of the electron which Lorentz had introduced. One of the first models was proposed by Sir. J. J. Thomson. He conceived of the chemical element as a positively charged sphere in which the negatively charged electrons are imbedded in static equilibrium. But this model failed to account for the extreme deflections of alpha particles which occur when they are projected through very thin sheets of metal. Hence Rutherford was led to introduce an entirely different conception. The positive part of the atom was regarded as compressed to a very small relative portion of the whole, and its total charge was regarded as made up of a number of elementary charges. About this positive center, or nucleus, the negatively charged electrons, equal in number to the atomic number, moved in planetary orbits. In this fashion the dynamic theory of the chemical atom arose. Since the mass of this atom is concentrated mostly in the relatively small nucleus, it follows that the mass of matter is mostly inner field. This dynamic model had the merit of explaining the radical deflection of alpha particles, but before other facts it failed utterly. Since the system must radiate energy continuously with the continuous rotation of the electron in its orbit, it followed that the frequency of the emitted light should decrease accordingly. Such a result is out of accord with the findings of spectroscopy.

Into this situation came the young Danish physicist, Neils Bohr.[4] With a single stroke he united the ideas of Planck, Einstein, and Rutherford to produce a new model of the chemical atom and a new mechanics. Taking Rutherford's dynamic atom, he assumed that there are only a few orbits or stable states in which a given electron can move, and that these states are related to each other by a certain rule which involves Planck's constant. Furthermore, the atom does not radiate when the electron moves continuously in these states. Instead, radiation occurs only when the electron jumps from one of them to another. Moreover, the change in the energy-state of the atom, which results from such a jump, is always in terms of a discrete unit or quantum of energy. When the electron jumps inward from one permissible orbit to another the atom loses a quantum of energy which appears in the form of emitted radiation. When on the other hand, light falls on a body a certain quantum of energy is absorbed and the electron jumps to an outer orbit. In this fashion Einstein's idea of the quantum of radiation is joined to Planck's theory of the quantum of energy, and to Rutherford's theory of the dynamic atom, and a verified connection between the electronic structure of the simplest chemical atom and the type of spectral lines which it emits is established. This unification exhibits itself in the following equation:

$$\nu = \frac{W_1 - W_2}{h}$$

Where ν is the frequency of the emitted light, W_1 and W_2, the energy in the quantum orbit from which and to which the electron jumps, respectively, and h is Planck's constant.

The genius of Bohr is not to be underestimated. His relation to quantum mechanics is similar to Newton's with reference to classical mechanics. As Newton unified the separately discovered ideas of Galilei, Kepler, and Copernicus, so Bohr brought together the ideas of Planck, Einstein, and Rutherford. The immediate verifications were equally convincing in both cases. The hydrogen spectrum and the periodic table of the chemical elements took on new meaning in the light of Bohr's theory. A definite relationship seemed to be established between the electronic structure of an atom, its chemical properties, and the type of spectrum which it emits. Moreover, the relation between frequency, energy states, and Planck's constant which Bohr introduced carries over into all the later theories. But time was not as lenient with Bohr as with Newton. For Newton could die and two centuries could pass before shortcomings in his theory were established, but Bohr did not even succeed in getting the full consequences of his brilliant synthesis into writing before inadequacies were evident. So great is the flux and change in this branch of science.

For chemical elements of high atomic number, the theory did not work. Also, the laws of dispersion, among other things, refused to give way to its attack. Thus scientists were finally forced to the conclusion that the difficulties involved in his theory were not merely practical in character, but theoretical. The theory itself is in some sense inadequate.

MATHEMATICAL THEORIES

The first attempt to develop a theory which can deal with the difficult problem of dispersion was made by Kramers.[5] He threw out the classically defined orbits of Bohr's theory, and was guided by Bohr's quantum correspondence principle alone. The results were successful.

This suggested to Heisenberg that the physical model could be ignored and only the mathematical equation considered. So the latter urged that no quantities be introduced into one's equations which cannot be physically observed. Science must restrict itself to measured quantities and mathematical relations. The previous demand for physical models was referred to as an unjustified desire to think in terms of images. Thus what appeared as a local necessary evil was turned into a virtue and made a philosophy. It did not matter if more certain branches of science gave no evidence for the validity of such a doctrine. The emphasis of the moment did, and that was enough. Place one of our contemporary physicists in a New England trout stream, and he could easily convince himself that reality is nothing but water, so great is his capacity to concentrate upon that which is most immediately observable at the moment. The fact that his rod is made of solid matter and that he is standing upon a rock-bottomed brook would enter into his philosophy no more than the fact enters into his physics that his measured quantities are determined with rods and clocks which are physical objects and that the experiments which his science performs are meaningless unless the physical theory of nature is assumed. The desire for physical models does not have its basis, as so many of our contemporary scientists assume, in the perverse tendency of the human mind to think in terms of images, but in the rationale of all modern and contemporary experimental scientific procedure. It must never be forgotten that the experimental method of physical science did not arise until science accepted the physical theory of nature. If nature is nothing but a group of sense data then experimental procedure is pointless; it is only if nature is a larger system of masses and motions of which the local experimental apparatus is an essential part that the performing of experiments is of any theoretical significance. It is not the contemporary experimentalist who insists on physical models who is inconsistent, but the contemporary theoretical physicist who thinks that one can have experimental physics without accepting the physical theory of nature.

But it is not necessary to develop this point in detail. For even Heisenberg himself found it impossible to carry his philosophy through, even in his own restricted field. Certain developments occur before this happens, which must be considered. Moreover, the procedure which Heisenberg initiates is fruitful, notwithstanding the fact that it cannot be generalized into a complete philosophy of science. It often happens in the physical theory of nature, that one must pay attention first only to relations and leave physical foundations to a later time when more information is at hand.

In any event, when Heisenberg proceeded to develop quantum theory in this fashion, certain complicated equations followed. At this point Born [5] made the important discovery that these equations have precisely the same form as those involved in a branch of pure algebra known as the multiplication of matrices. The suggestion was obvious: Let us build atomic physics upon the principles of these algebraic forms. This was done by Born, Heisenberg, and Jordan. The result is a conceptual formulation of quantum theory, known as the matrix theory. It is not to be overlooked, however, that the fundamental discontinuous ideas of Planck still carry over.

But this extremely mathematical type of formulation did not stop at this point. Dirac made an additional generalization, introducing steps which cannot be represented in matrices, and using more general ideas from which the theory of matrices follows as a special case. Born also proceeded in the same direction. Suffice it to say that in conjunction with Wiener, he is led to the notion of functional operators, in order to find a meaning for quantum magnitudes, in the case in which the theory of matrices fails.[5] This new conception is to be noted merely as a signpost which marks the point in this extremely conceptual branch of quantum theory at which an independent development of a quite different character appeared and was most fruitful. This brings us to wave mechanics and the ideas of L. de Broglie[5] and Schrödinger.[6]

WAVE MECHANICS

Before turning to these new conceptions, it is important to note that the generalizations of Heisenberg, Born, and Dirac involve very much more than a mere play with an obscure branch of mathematics; they also give rise to laws which fit given optical and spectroscopic evidence with a degree of accuracy and completeness which previous theories have not attained. All these extremely mathematical formulations bespeak something which has a basis in the nature of things. Precisely what this is, is not clear as yet in all its details, but a negative consequence is already evident and can be stated now. The attempt to develop a theory of the chemical atom and its behavior in radiation, by conceiving of it as built up out of nothing but its constituent parts has broken down. Even in the Bohr atom the signs of this result were evident, for the orbits represented restrictions placed upon the supposedly possible motions of the constituent particles; they were not paths definable in terms of the intrinsic properties of the particles themselves. One additional fact may be noted. The character and behavior of the chemical atom is not to be separated from its radiation. In other words absorption and radiation are not effects from, or upon, other external and isolated atoms; but atoms and neighboring atoms and radiation are all parts of one single system. Since the traditional physical theory of nature provided no basis for such a conception, it was natural that atomic physics should be forced to an extremely mathematical formulation in order to be able to deal with the facts.

But if nature is a physical system it should be natural also for further study to force even this branch of science to a reaction against such a development, which would exhibit itself in a new attempt at a physical theory. This is the actual case. It appears negatively in the failure of Heisenberg, who is the author of this conceptual type of procedure, to maintain his own philosophy; [8] and positively in a remarkable discovery by De Broglie and its extension and application by Schrödinger.

Let us consider the negative side of the picture first. In this connection mention must be made of a very important discovery by Compton. He found that when light is radiated from a given atom and strikes another chemical substance in which the electron-proton structure is not very rigidly fixed, the light wave can cause a given electron to move off in space. Moreover, the speed with which the electron departs is independent of the intensity of the light. Also, in this process the wave itself suffers no decrease in velocity, but merely a decrease in frequency. Thus waves behave as if they were corpuscles. This is known as the Compton effect.[8]

Heisenberg noted that this throws new light upon the scientific doctrine of causality. Mechanical causation may be defined as the thesis that a knowledge of present phenomena will enable one to predict all future effects. It has been supposed that modern science accepts this doctrine. Heisenberg points out that the Compton effect makes it impossible for us to determine both the velocity and position of an electron at the same time. For one cannot determine the location of an electron without bringing light to bear upon it, and the effect of this is to put it out of its position. Thus the more accurately one tries to observe, the more inaccurate are the results of one's observation. Hence Heisenberg concludes that it is impossible for us to observe the initial conditions of natural processes. This doctrine is known as the principle of indeterminism. It has been generalized and is now known as the uncertainty relation.[8] The suggestion arises, therefore, that there is contingency at the basis of things.

Two points are to be noted with reference to this doctrine. First, it rests upon Heisenberg's philosophy, and second, it denies that philosophy. The first point becomes evident the moment one notes that this argument for indeterminism does not hold unless one accepts his doctrine that the only meaning a scientific concept can have is one given in an experimental operation. Indeterminism holds, he says, because it is experimentally impossible to determine the conditions for determinism. In other words, the only meaning which a scientific concept may have is one defined by a technical experiment. But science has other ways of arriving at meanings for its many conceptions. The method of hypothesis is one of them; the statistical method, a second; and the procedure of the Greek inductive natural philosophers, a third. Thus while Heisenberg's argument leaves science open to a doctrine of indeterminism it cannot be taken too seriously, since it is impossible to define the accepted methods that science has used in its history in terms of the restricted philosophy upon which it rests.

This becomes evident in this very instance. For Heisenberg only establishes his principle of indeterminism by assuming electrons and their motion. These conceptions involve not merely a physical theory of nature, but the rejection of the philosophy which makes his doctrine of indeterminism plausible. This philosophy permits the introduction of no factors aside from mathematical formulae which are not immediately observed. Electrons do not satisfy this requirement. The fact that he finds himself forced to use them in defining one of his most important conceptions, indicates that the attempt to develop quantum theory in purely conceptual terms is breaking down. Since the above was written, a book by Heisenberg[8] has appeared in which he frankly admits that the experimental verification of every scientific concept is an impossible ideal, and in which he returns to physical conceptions.

The positive side of this development appears with L. de Broglie. He notes that particles behave as if they were waves, and waves as if they were particles. Neither the traditional continuous nor the traditional discontinuous theory will suffice. The difficulty is not that we must give up thinking in physical terms but that we have the wrong physical conception. Out of these reflections a new hypothesis arises. De Broglie writes[5] "I shall suppose that there is reason to admit the existence, in a wave, of points where energy is concentrated, of very small corpuscles whose motion is so intimately connected with the displacement of the wave that a knowledge of the laws regulating one of these motions is equivalent to a knowledge of the laws regulating the other. Conversely, I shall suppose that there is reason to associate wave propagation with the motion of all the kinds of corpuscles whose existence has been revealed to us by experiment." With this doctrine that any particle is associated with a wave in a determinate fashion, he is able to take the laws of wave propagation as fundamental, and to deduce the laws of a dynamic particle from them. In this manner, the new science of wave mechanics arises.

The important point to note in connection with this theory is that atomic motion and propagation are inseparably bound together. The traditional procedure of starting with constituent microscopic atomic parts and building up chemical elements and their interactions out of them has failed. An adequate theory of atomic and optical phenomena must conceive of them as a single system in which the propagated light determines the motion and behaviour of the particles as much as the particles determine the nature of radiation. Physical science has come upon the same fact in atomic theory that appeared in relativity theory. Just as the demands of a physically conditioned metric for measuring, forced science away from the doctrine that all relatedness between physical objects is a variable contingent effect of microscopic atomic motion, so the optical behavior of chemical elements is forcing physics to the conclusion that the relations between atoms, which exhibit themselves in electro-magnetic propagation, cannot be regarded as the mere effects of a simple compounding of electrons and protons. The wave or macroscopic aspect and the corpuscular or microscopic aspect are both parts of a single system, and neither can be defined in terms of the other.

Two events forced science to take De Broglie's ideas seriously. First, Davisson and Germer [8] in America and G. P. Thomson [7] in England demonstrated their validity experimentally. Thomson's experiment revealed that a stream of fast electrons which are passed through a thin film of metal give rise to a wave formulation, the measurable quantitative characteristics of which are in precise agreement with the theoretical rule which De Broglie laid down. Second, Schrödinger [6] developed De Broglie's conception mathematically introducing imaginary dimensions, and showed that it not merely provides a meaning for the Born and Wiener 'operators' in terms of which the matrix theory had been expressed,[5] but also brings order into this realm where chaos had appeared before.

To leave the impression, however, that everything is in a satisfactory state, or that current wave mechanics is unequivocal in favor of a physical theory of nature would be misleading. Schrödinger's original physical conception of his work has had to be modified at least in part. Many students in this field will maintain that this modification has gone so far as to leave the purely conceptualistic philosophy the orthodox opinion. The plain fact is that changes have come so fast, and the subject-matter is so complicated, that the full force of what has happened is not grasped, and the demands of an adequate theory are not yet realized. At such times emphases of the moment are likely to be given more weight than they deserve and the commonplaces that remain constant through all the changes are overlooked.

A few elemental points may be noted, however. In the first place, Planck's constant and the relation, by means of which Bohr connected the energy states or wave formation of the atom with the frequency of the propagated light by means of this constant, still stands. This means that nature must be discontinuous in a fundamental sense, for these ideas entail, as we have noted, a discontinuous theory of energy and a discontinuous or finite theory of the units in terms of which randomness in energy distribution is defined. Any theory which retains Planck's constant is atomic in some fundamental sense. In the second place, none of those who have attempted to develop purely conceptualistic or continuous theories, have succeeded in making such a philosophy work. Heisenberg has been forced to state one of his most important contributions to contemporary scientific thought in terms of the electron and its motion; and the notion of a particle is an essential notion, both in theory and in practice, in wave mechanics. Moreover, something more than the microscopic particles of the traditional physical theory must exist. The reasons for this conclusion need to be repeated. The stable states of the Bohr atom which place limitations upon the motion of the microscopic particles were the first evidence of this. Here was something elemental and purely formal which not only is not physical, but which conditions the motion of the physical. The breakdown of the Bohr atom when one considered systems more complicated than the hydrogen atom, the appearance of the wave as well as the particle as fundamental in De Broglie's theory, and the original purely formal or mathematical theories of Heisenberg and Born are but extensions of this fact. There is something in the realm of atomic physics, as it bears on the propagation of light, which refuses to be resolved analytically into nothing but the microscopic atomic particles built up into more complex structures as bricks are added together to make a house. Briefly, the purely analytical approach to nature has broken down. The relation of the microscopic atom to light and to its neighboring microscopic particles is as fundamental as the particle itself; the one conditions, and is conditioned by, the other. Field or macroscopic as well as atomic causes are present. Wave mechanics with its specifications of boundary conditions is an expression of this fact. These points may be summarized by saying that current quantum and wave mechanics necessitates a discontinuous theory, but also reveals a residual continuous and organic or formal mathematical factor, which our traditional physical theory of first principles is incapable of rendering intelligible. This suggests that a modification in our traditional physical theory of nature is demanded. May it not be fruitful, therefore, to go back to the first principles, which our analysis of the theory of relativity has established, and proceed into quantum and wave mechanics from this direction. When this is done the recent developments which we have just outlined are not as strange, or as devoid of physical significance, as their first appearance suggests.

EMPIRICAL EVIDENCE IN THE LIGHT OF FIRST PRINCIPLES

A determination of the necessary consequences of the theory of relativity led us to the discovery of a new theory of the first principles of science. This theory conceives of nature as made up of a finite number of moving microscopic particles surrounded by one spherical macroscopic atom. The precise nature of the moving microscopic particles can be determined only by intensive analysis in the technical sciences. Available evidence suggests that the electron which is the current microscopic atom, is not merely the concentrated bit of stuff which is designated as its central charge, but also a field radiating out from that charge. Thus, it has been said that the electron is everywhere; only its central charge is in a certain place. In acordance with this conception, the macroscopic atom is also to be conceived as not merely a surrounding hollow sphere of indivisible material, but also an inner field. All these atoms obey the principle of being; that is, neither their material nor their motion is created nor destroyed. This is but a more explicit way of saying that our theory is a physical theory. When the properties of such a physical system are specified certain interesting consequences with reference to optics, and quantum and wave mechanics appear.

The macroscopic atom is finite in size and is so small, relatively to the finite number of moving microscopic atoms, that it congests them sufficiently to produce a general uniform and approximately constant compound structure over macroscopic distances, but not so completely as to prevent the motion of the microscopic atoms and the resultant variable relatedness which their inherent motion introduces. We have noted that the theory of relativity necessitates this conception. For it reveals that precisely such a metric exists and that it is conditioned by matter.

It may be noted at the very outset that a fundamental principle of polarity or duality appears in the metaphysics of the kinetic atomic physical theory of nature with the introduction of the macroscopic atom. This atom provides the principle of rest, the microscopic particles the principle of motion; it introduces the principle of unity, they, the principle of plurality. Likewise its shape entails the principle of necessary and fixed form, their motion the principle of variable and contingent relatedness. Thus in terms of a polar physical system we derive a metaphysics with its thesis, antithesis, and synthesis which combines the universality of the Hegelian dialectic, with more fertility in the concrete sciences, and the precision which only an analytical and pluralistic system can give. If everything is unity and the thesis and antithesis are nothing but abstractions, and all is relative and all is absolute from different standpoints, thought is always muddy and ambiguous because it is always shifting its premises to suit the occasion. Because of this Greek thought as it was later developed along monistic lines lost the beauty and clean-cut clarity which was its essential and original character. In this respect at least, the macroscopic atomic theory is a return to the pure Greek tradition. It is basically and unequivocally pluralistic. Its macroscopic atom stands in a logically external relationship to the microscopic atoms, and the latter in turn enjoy the same type of relationship to each other. Moreover, the microscopic atoms possess inherent motion; but it cannot occur without the macroscopic atom. This is a point which must be watched. It may have very important implications.

The difference of property between the microscopic atoms and the macroscopic atom introduces the required principle of polarity and duality which the whole of philosophical analysis from Plato's "Parmenides" and "Sophist" [10] through Hegel's "thesis" and "antithesis" [11] to Cohen's "principle of polarity" [12] and Sheldon's "principle of productive duality" [13] has revealed to be necessary. But many metaphysical systems have done this before. The uniqueness of ours is that it is accomplished for the first time with the physical theory of nature, which modern science has revealed to be necessary, and by providing a meaning for the synthesis of the opposites without forcing one to monism and the doctrine that all analytically gained concepts are abstractions. This happens because the spherical encircling property of the macroscopic atom, which provides the antithesis to any theses introduced by the microscopic particles, also necessitates a synthesis of these opposites.

This synthesis exhibits itself first in the production of a compound field. It is this compound field which so many are mistaking for an irreducible space or ether. All field equations refer to it. The number of microscopic atoms and their acompanying fields is so large, compared to the finite character of the macroscopic atom and its inner field, that compression and combination must result. The microscopic atoms are in motion. Hence they tend to produce a changing compound field. The macroscopic atom is changeless and of a fixed form. It tends to produce a changeless compound field. But it also necessitates that the variable structure produced by the microscopic motion, must merge with the static type of field which it introduces. The synthesis must be a compound field or ether with macroscopic metrical uniformity and constancy interspersed with microscopic heterogeneity and variability, such as relativity physics has revealed.

Furthermore this compound field will contain two factors. There will be compounds of the fields of the basic atoms, and there will be compounds of the central charges of these fields. The latter, science calls chemical atoms, molecules, molar bodies and nebulae; the former, the region in which molar bodies are imbedded. It is for one equation, describing both the regional and discontinuous characteristics of the compound field, that Einstein is now seeking. He has been inclined, in his recent more popular utterances,[14] to refer to this compound field as space. It seems more in accord with modern mathematical usage to reserve the term space for relations between physical objects, as Einstein did in his earlier writings, rather than to regard space as an ultimate, which, to use his recent language, is "swallowing up the field and the particles." [14] Moreover the latter attempt to derive the discontinuous from a continuum can but lead to insuperable difficulties with reference to motion. Only if the continuum is regarded as a compound physical field finding its basis in a polar principle, according to which macroscopic unity is merged with microscopic discontinuity and motion, can these difficulties be avoided.

The more general field, which is made up of molar objects and their intervening regions, may be termed the ether. Since such an ether is variable in structure, and is defined in terms of matter, and is constituted in part by our earth, it follows that ether drift experiments are irrelevant so far as the validity of the theory of relativity is concerned. It is only a non-molar static ether independent of molar motion that is inconsistent with Einstein's theory. In this fashion the principle of polarity and synthesis of the macroscopic atomic theory, not merely meets the metrical demands of relativity physics, but also provides electro-dynamics with a physical medium for the transmission of its waves. We shall find, however, that our ideas of the nature of light must be radically modified.

But there is not only a polarity of fields and substances but also a polarity of general properties. In fact, if the macroscopic atomic theory is true, nature must be a combination or mixture of order and disorder, of uniformity and variability, of permanence and change due to motion, of necessary and contingent relatedness, and of continuity and discontinuity. Thus a principle of opposition and contradiction is inherent in nature, and any principle except that which defines the macroscopic atomic theory, gives rise to its negate when pressed far enough.

In this fashion our theory would provide a meaning for the indeterminism which Heisenberg has suggested, without running the risk of making the existence of science a mystery and a rational account of natural processes impossible. Were the only relatedness in nature, the product of the motion of the microscopic atoms, scientific laws which refer to relations would be impossible; the chaotic aspect of experience would be the general rule. The general macroscopic uniformity and constancy which the macroscopic atom imposes prevents such devastating consequences; it adds an element of necessity to the order of nature. Thus for the first time, the physical theory of nature is able to do justice to the fact of form and order in nature and its parts, and, at the same time, account for the element of disorder and disintegration which also exists. All relatedness is not necessary, but neither is it all contingent, as the existence of geometry and the metrical uniformity and constancy involved in measurement eloquently testify. Instead, the relational order of nature is a mixture of the necessary and the contingent.

But even a partial amount of elemental contingency is unthinkable and inexpressible. As Socrates indicated long ago, unless the principle of being holds even conversation is impossible. Our new physical philosophy meets this difficulty, since, unlike most philosophical systems, and particularly contemporary ones, it gives rise to two systems of logic. The atoms, as independent entities, are substances with properties. This aspect of them is fundamental. Thus they provide a basic meaning for the subject-predicate logic of Aristotle. But the motion of the microscopic particles and the fixed form of the macroscopic atom gives rise to two opposed types of relatedness between their compounds. The relatedness produced by the macroscopic atom is necessary in the sense of being changeless; that produced by the moving microscopic particles contingent, since it varies. This occurs because the form of the macroscopic atom is changeless, whereas the relational forms of the microscopic particles are changing, because of their motion. Thus a logic of relations arises. It is in the realm of this logic that contingency occurs, and that a meaning for possibility is found. It is here also, as we shall indicate later, that contradictions find their basis, and that a meaning for the distinction between the necessary and the possible is to be discovered. This fact exhibits itself unequivocally in the history of science. There are a large number of possible geometries. Thus, a meaning for the possibles in the logic of relations is defined in terms of that which is necessary in the logic of Aristotle. A meaning for contingency in the sense of variable relatedness exists and nevertheless we can talk about it. Such cannot be the case in a philosophy which rests on pure contingency.

The opposition between relational necessity and relational contingency, which the macroscopic atomic theory necessitates, exhibits its fertility in quantum and wave mechanics. The motion of the microscopic particles introduces relational variability. Something other than them must exist, therefore, to provide the uniformity which exhibits itself in the orbits of physical motion and in the laws of natural science. The fixed form of the macroscopic atom introduces this additional factor. Hence it follows that the first complex substance, to compound out of the basic atomic entities of the macroscopic atomic theory must exhibit this peculiar combination of the contingent relatedness of the microscopic and the necessary relatedness of the macroscopic. If our theory is true we should expect to find this union of contingent and necessary relatedness in the structure of the chemical atom.

Needless to say, Bohr found precisely this. There are stable states and there are breaks, unpredictable except on statistical grounds, from those stable states. Again the polar principle in our philosophy exhibits its existence. Since the only chemical properties of molecular and molar bodies which traditional science considered were those which remained constant, or practically so, through the contingency which the jumps of electrons introduced, our failure to find contingent relatedness before the time of Bohr is explained. It was possible, therefore, up to his time to imagine that one can build science without the polar opposition of necessary and contingent relatedness which the macroscopic atomic theory introduces. Since his time, however, atomic physics and optics have exhibited a series of oppositions. At last science has touched the foundations of compound bodies and systems where the true sources of natural order and stability exhibit themselves. The result is a complete revision in our most elementary physical conceptions. Compound bodies such as the chemical atom are not built up from within by internal attraction out of microscopic particles alone, but owe their stability instead to the opposition between a local microscopic tendency to break from stability and an externally imposed tendency to preserve stability.

For this reason, the stability of the atom is part and parcel of the wider stability of nature as a whole. This means that an atom cannot reorganize its structure without the whole universe reorganizing itself. Nothing is built up in piece-meal. This is not theory, but verified fact; for we have evidence of the reorganization of nature which accompanies every local chemical reorganization. It appears as light. In fact light may be defined as the thesis that an electron cannot break from its position as one of the basic foundations of the structure of the universe without the whole universe reorganizing itself. Is it any wonder that physical chemistry and optics are inseparably bound together?

It is as if all the bricks in a building were continually shifting, and all were so organically related to each other that a change in the position of any one, entailed an alteration in the structure of the building as a whole. Obviously such shaky foundations would produce a succession of reorganizations in the structure of the whole. Our universe is of this character. For it is an equilibrium between the fixed outward pressure of the moving microscopic particles and the fixed inward pressure of the one macroscopic atom. Since the properties of both are fixed, the mean result must be a constant. Hence any reorganization of a chemical element or of nature as a whole must obey this constant. Physics has actually found this constant. It is known as Planck's constant and designated by the letter h. One theory gives it a physical meaning, and a theoretical justification for the first time. It is to be defined as a function of the diameter of the macroscopic atom, and the number of microscopic atoms which it contains.

But this is not all. The microscopic foundations of this equilibrium are in motion. Moreover, the general tendency of their motion is rectilinear, whereas the general tendency of the motion which the macroscopic atom is introducing is circular. Hence an opposition of these two motions results in the continual breaks from equilibrium which Bohr has postulated. But a structure under constant pressure must reorganize itself when one of its foundations gives way. Hence there will not only be a shift in the atom but also a shift in the whole structure of nature. Moreover, this reorganization must be accompanied by a release of the potential energy which opposing forces store in the stable equilibrium. Hence when an electron jumps, energy will be released. Were this break from equilibrium not compensated, the whole universe would go to pieces. But because of the spherical character of the macroscopic atom, an electron only breaks from the stability which the macroscopic atom imposes in one direction, by running into it another. Hence the releases of energy are finite in quantity and come in spurts, and nature, on the electronic level, is a series of breaks from one reorganization to another, rather than a sudden and complete disorganization. The character of the macroscopic atom is sufficient to insure stability, but not sufficient to bring the contingent radical erratic tendency of the microscopic particles completely into line. Hence finite spurts of energy are continually released, and energy has its quantum-like character in both physical chemistry and optics.

Moreover, any physical system is as much a field as a collection of microscopic particles. For the macroscopic atom produces a compounding of fields and preserves a certain macroscopic uniformity in them. Now, any change in a field which is an equilibrium of opposing forces must exhibit itself as a wave. Hence any moving particle will be accompanied by a wave. A reorganization of the compounds of the central charges of the microscopic fields will produce a reorganization in the ether, and a reorganization in the ether will entail a motion of the central charges of the local microscopic atoms which are its parts. Thus waves must behave as if they were particles and particles as if they were waves, and the attempt to build up a model of the chemical atom, that will work for its optical effects in emitted spectra, in terms of nothing but the microscopic particles must fail, as Heisenberg and Born and others have discovered. Current developments in atomic physics are precisely what one would expect if the macroscopic atomic theory is valid and scientists had not yet discovered the existence of the macroscopic atom and the polar theory of the stability of the chemical element and nature, which it introduces. It is hardly necessary to add that this peculiar combination of kinetic atomic and wave properties which any sub-atomic change must exhibit, if the macroscopic atomic theory is true, has been suggested by De Broglie, developed by Schrödinger, and discovered experimentally by Davisson, Germer, and G. P. Thomson. We may summarize this point by saying that wave mechanics is a discovery of the fact that both the compound ether and compound physical objects are constituted not merely of the microscopic atoms which compose them from within, but also of the field formulation and polar equilibrium which the macroscopic atom imposes upon them from without.

Another important consequence follows if the macroscopic atomic theory is true. The microscopic particles tend to produce disorganization; the macroscopic atom organization. It follows therefore that our universe must be a mixture of macroscopic order and microscopic disorder. Moreover, since the properties of the basic atoms are fixed and changeless for eternity, it follows that the amount of organization in nature must be a constant. Hence if a chemical atom disorganizes in one place, another chemical atom or complex substance must be built up somewhere else. Stated in terms of energy this means that the quantum of energy emitted by the jump of an electron in one molecule must be balanced by the absorption of it somewhere else. Thus the emission and absorption of light is a symmetrical process. G. N. Lewis has expressed this point by saying that an electron cannot jump unless it knows that the energy which it emits will be properly received somewhere else.[15] We may add that the macroscopic atom, because of the constant total amount of organization which it necessitates, guarantees to any chemical element that it can release its energy at any time without any risk of being snubbed by its neighbors for this uninvited burst of generosity. This fact, Lewis has termed the principle of virtual contact. He deduced it from the theory of relativity.

These conceptions need to be brought into conjunction with the doctrine of the primacy of eternity over temporality, to which we referred at the end of the last chapter. We deduced the latter principle from the fact that the theory of relativity makes motion and propagation primary and defines space and time in terms of it. It appears, however, in a more primary form in the principle of being,—the thesis that the basic physical entities do not change their properties, and in the corollary that the motion of the microscopic particles is not in space or time, but in the changeless macroscopic atom. Thus time and space appear as changing relations between physical entities, and not as primary concepts. This notion, that motion and light propagation simply *are,* must be grasped if the doctrine of the same velocity of light for all frames of reference, and Lewis's principle of virtual contact, and the breakdown of spatial and temporal conceptions in quantum and wave mechanics are not to be absurdities. At no point does the need for an adequate philosophy, if we are to understand our physics, make itself more evident than here. It is only because nature permits us to regard motion and light propagation as elemental and hence outside time and space that the theory of relativity and the new atomic physics become intelligible, and certain of their consequences, such as Lewis's principle of virtual contact, make sense. The macroscopic atomic interpretation of quantum mechanics and radiation adds to the intelligibility of these conceptions. For it reveals that the jump of an electron and the propagation of light is not a serial temporal process but a reorganization of the structure of nature as a whole. Hence light propagation so-called, simply is. Its linear character is physical and logical rather than temporal. Hence in a given relative frame, when measurements are made and everything must be given relative space and time values, the numerical value for the relative spatio-temporal expression of light propagation may be the same for all frames.

It cannot be too strongly emphasized that the primary notion which one must grasp, if current scientific discoveries are to be understood is the supposedly outworn doctrine of Greek science, that the notion of eternity is more fundamental than the notion of temporality. Physics, the science which knows more than any other about time and its relation to the elements of nature, has revealed that the temporal and evolutionary aspect of nature is most superficial and relative and can provide no basis for a sound theory of first principles. No theory is more incompatible with contemporary physics than the doctrine of emergent evolution.

The manner in which the macroscopic atomic theory provides a consistent and physical meaning for the unusual ideas of quantum and wave mechanics constitutes another argument for its validity. By accepting it, relativity and quantum mechanics are brought into relationship with each other. Only at one point does our theory give the lie to an accepted doctrine of current science. It is incompatible with the thesis that the second law of thermo-dynamics applies to the whole of nature. This law represents but the microscopic tendency of our theory. The effect of the microscopic particles is to produce a breakdown of differentiated structure and a tendency to homogeneous disorder and chaos. But the macroscopic atom prevents this. Its finite size imposes an organization upon nature which is necessary. Hence, as we have indicated the mean between the organizing and the disorganizing tendency must be a constant for the whole of nature, and the amount of organization in nature as a whole must be invariant. To be sure there is nothing to prevent a long process of disorganization occurring in one place providing an equivalent process of organization occurs in another.

Only if the microscopic atomic entities are the only constituents of nature can the second law of thermo-dynamics apply to nature as a whole. But the crucial question still remains. Does the second law of thermo-dynamics which Sadi Carnot proposed, and which modern science has confirmed again and again, permit of the interpretation which our theory presents? This entails an analysis of the second law on its own merits.

THE SECOND LAW OF THERMO-DYNAMICS

This is one of the most interesting and baffling of the laws of science. It was discovered in 1824, without any forewarning of its coming, when a young Frenchman, Count Sadi Carnot, reflected on the efficiency of the heat engine. The law has two interpretations, the one literal, the other, statistical. The latter is the current orthodox conception. We shall consider the former meaning first.

Carnot noted that one can get work out of a given system, which involves a transition of thermal into mechanical energy, only with a drop in temperature. The steam engine provides for this in its actual anatomy with its boiler and condenser. The operation of the principle appears also, in our incapacity to make use of the tremendous store of energy in the sea, because its temperature is lower than that of any available system in our environment. The tendency of the temperature of any system to become uniform with that of its environment is another example of our rule. Stated in general terms, this means that work can be accomplished in our universe only by removing differences in temperature. It seems natural to conclude, therefore, that the changes which occur in nature are hastening the day when no difference of temperature exists and no work can be done.

Moreover, since organized systems call for work to be done upon them, and the tendency to uniform temperature represents an increasing incapacity to do work, it seems evident that the second law entails a tendency to homogeneous disorganization of structure, as well as to uniformity of temperature. In short, we are proceeding to frigid chaos and dust.

In 1842, eighteen years after the discovery of the second law, a German physician, Mayer, announced the first law of thermo-dynamics,—the doctrine that energy is neither created nor destroyed. This discovery was made secure when Joule determined the mechanical equivalent of heat. At first, the principle of Carnot seemed to be inconsistent with these findings. But it was soon noted that the second law must be true, if the law of the conservation of energy is to be reconciled with the impossibility of perpetual motion in local systems. If energy is not destroyed, why is it impossible to produce a machine which will run and do work without fuel? The answer comes when one becomes aware of the distinction between energy that is available for work and energy in an unavailable form. The first law asserts that the sum of these two types of energy is always a constant; the second law, that the amount of energy which is available for work is decreasing while that in an unavailable form is increasing. A local machine which will run without fuel is impossible, because work can be accomplished only with energy which is available for work; and because the potential energy which we put in our machines as fuel, can be transformed over into an active kinetic form only by producing a certain amount of degraded heat which is unavailable for work. It is to be noted, therefore, that the principle of Carnot, is not a doctrine which can be idly brushed aside to suit our ethical aspirations; it has a very real validity in the experience of each one of us.

It also finds justification in the traditional first principles of modern science. According to these principles nature is merely a collection of moving microscopic particles. In fact, Count Rumford's experiments in connection with the boring of a cannon had led him to the conclusion that heat is disorganized molecular motion. This made the discovery of the mechanical equivalent of heat intelligible, as it also accounted for the verified physical laws relating the volume, pressure, and temperature of any gases; laws which Boyle, Marriotte, Gay-Lussac, and Avagadro had discovered. One has but to regard energy which is available for work, as organized molecular motion to complete the picture and derive the second law. For if nature is at bottom nothing but moving microscopic particles, the variable relatedness which they produce must eventually give disorganization a victory over organization. This consequence becomes the more precise, if one recalls a distinction between potential and kinetic energy which classical science had introduced in order to save the first law. Potential energy is energy due to pressure or organization and position; kinetic energy, as its name indicates, is energy due to motion. Strictly speaking, the notion of potential energy had no theoretical justification in classical science. At bottom only kinetic energy existed. But in fact, many examples of potential energy were present: the coal before it is heated, the crude unsparked oil, the stone on the shelf, or the water at the top of the precipice,—all these are cases of energy in a potential form.

But potential energy cannot do work until it is released. The release can be accomplished, however, only by destroying the existing organization of molecules. It follows quite naturally that such disorganization will allow certain particles to escape in chaotic motion, and permit only a portion of the original molecules to move in a newly directed and organized fashion to do work. Thus work is done only by decreasing the available amount of organization in nature, and increasing the amount of disorganized molecular motion, which is termed dissipated heat.

It is to be noted, therefore, that the second law of thermo-dynamics owes a great deal of the authority which it has enjoyed, to our traditional theory of first principles. In fact, it could have been deduced from those principles had Sadi Carnot and the empirical science of thermodynamics never have come into existence. Certainly, the principles governing the efficiency of the steam engine, and even the local confirmations and reconfirmations of modern physical chemistry alone, are not enough. For science has learned again and again of the danger of inferring from a local instance, or even a local many or all, to a universal all. Since we have already discovered that the first principles of traditional science, which have justified this generalization, are incomplete, we must be somewhat skeptical concerning the applicability of the second law to the whole of nature.

The probability that our universe is finite accentuates this point. The problem of the stability of our universe, which led Einstein to his theory of the finite universe involves precisely the same question. In fact, it is very strange that the bearing of the theory of the finite universe upon the second law has not received more attention. Certainly, to assert that the spatial dimensions of nature are closed, so that a given physical entity cannot escape from the rest of matter, is to maintain that such thoroughgoing physical disorganization, as we have previously assumed, is impossible. We shall return to this point later.

But it is not necessary to appeal to first principles in order to become skeptical concerning the universal applicability of the second law. Two obvious facts raise serious difficulties.

In the first place, there is the existing organization of our universe. If the microscopic kinetic principle is the sole factor at the basis of things why is it here? If the reply is that nature is on its way to chaos, then why are we not there now? To say that the process is an infinite one is no answer, for it follows from the nature of infinity that our universe has had an infinite amount of time at its disposal already. Moreover, assuming this hypothesis to be true, another one also has to be introduced, which only increases the difficulty. To say that nature is running from an original state of highly complicated organization to chaos, is to raise the question concerning how the original state arose. Certainly it presupposes design, which in turn, entails a Designer. This is inconsistent with everything that modern science indicates concerning the first principles at the basis of things. Certainly the kinetic atomic theory with its emphasis on the principle of being and upon rearrangement, rather than essential creation, gives no justification for such a doctrine. Nor can the student of religion find any consolation in the suggestion of such a Designer. A God who would create a universe and then run away to allow it to proceed to chaos with rough-shod relentless destruction, through man and all his creations, can scarcely be the Heavenly Father of religion.

A more recent attempt to account for the original state of organization has been proposed by Jeans.[17] He suggests that matter is being poured into our universe from another dimension in space. Obviously relativity physics is supposed to make such a doctrine reasonable. Before such a proposal it is difficult to write with calmness. The plain truth is that this means absolutely nothing. In fact, it is worse than nothing since it is sheer nonsense. There is no objective space or space-time to provide, in one of its dimensions, for a trough along which matter can roll from some other universe, which has to be hypothesized, into this one. Only a person who has fallen into the error of denying the general theory of relativity by conceiving of space-time as an absolute background for matter could ever propose such a notion. But even if such conceptions were not ruled out by verified theories, would they be justified? When a reconciliation of known fact with a scientific theory necessitates such an interminable series of ad hoc hypotheses, is it not more scientific to take the known fact of organization at its face value, and question the universal application of one's theory.

When one considers the process of organic evolution the urge for such a procedure becomes the more emphatic. For organic evolution represents, not merely a present state of organization, but a movement in exactly the opposite direction to the one which the second law prescribes.

Instead of moving toward homogeneity of structure and temperature, organic evolution proceeds to increasing complexity and heterogeneity of organization, and diversity of temperature. Man is more complex than the lower forms of life which come at the beginning of the process.

However, one must not be too hasty in reaching a conclusion. The second law does not rule out the possibility of increasing complexity of organization. It is only for isolated systems that the tendency to uniformity of temperature is a necessity. Providing energy comes into a system from outside a building up process is possible. In the case of living things on this planet, this is not merely a possibility but a fact. We know that organic compounds are synthesized by the means of energy which comes from the sun.

But even with these facts, the universality of the second law is not saved without the introduction of an ad hoc and unverified hypothesis. It must also be assumed that the building up of organisms here is more than compensated by a breaking-down process somewhere else. Stated bluntly this means that the process of organic evolution is a mere incidental and accidental back-wash in an all-engulfing torrent of destruction.

Moreover, this hypothesis does not really meet the difficulty. It merely transfers it from biology to astronomy where it is hoped that it will be ignored. But difficulties are not met by moving them about from one science to another. The methods of the army are hardly those of physical science. Why, if astronomical nature is headed toward ruin, should it set into operation a process in exactly the opposite direction, upon such a grand scale as we find in organic evolution? This seems to be a very back-handed manner in which to proceed toward chaos and dust. Again we must query whether it is not a more sound procedure to take facts at their face value and admit the limited applicability of the second law, rather than to interpret and reinterpret facts with an endless series of most questionable unverified hypotheses, in order to save it in an unqualified form. May not the second law represent a logical rather than a temporal principle? In other words, may it not be but one basic principle in nature exhibiting itself in certain systems, to which another equally basic principle is opposed? Such a conclusion will enable us to account for everything that we know which is in support of the second law, and permit us to take other facts at their face value as well.

But it is not necessary to appeal to facts and face-value interpretations of them to justify this conclusion. It is involved in the present orthodox meaning of the second law itself. This brings us to the theory of probability and the statistical interpretation of the law.

Following Carnot, the German mathematical physicist, Clausius, presented a new statement of the second law. He asserted that for every state of an isolated system, there exists a function, which is determined by the nature of the system itself; and that this function, termed entropy, always *increases* in *one* and *only one* direction, with any changes in the system. Thus the second law became defined as the thesis that the entropy of the universe tends toward a maximum. The unidirectional character of this change means that there is a one-way process in nature. The state of maximum entropy which is its goal is the condition of homogeneous disorganization and uniform temperature to which we have previously referred. This may be termed the literal interpretation of the second law.

Later Boltzmann discovered that the entropy of a given state of a system can be connected to the probability of that state by a certain logarithmic expression.[18] This made it possible to define the second law as the thesis that change in any system is always toward its most probable state, which is the condition of maximum entropy. It is upon this interpretation that Willard Gibbs built his remarkable science of physical chemistry; and it is upon Gibbs' work, in turn, that the huge chemical industry of our contemporary industrial order reposes.

Moreover, the important thermo-dynamical conceptions of Planck, Einstein, and G. N. Lewis presuppose it. We may be sure, therefore, that any necessary consequences which we may derive from the statistical interpretation of the second law are reasonably secure.

It happens that such an interpretation entails certain very interesting and somewhat unexpected consequences. To make this clear it is necessary to consider the theory of probability as it bears on the nature of scientific law. Since this is a fascinating subject, on its own account, the necessity of having to examine it, somewhat in detail, is a pleasant one.

Let us suppose that we have a glass tumbler filled with fine grains of powder, and that all the grains in the upper half are black, and all in the lower portion white. Suppose also, as is the usual case, that we cannot see the individual grains. The total contents will appear as a half-black, half-white mass. We shall call this original state of the grains in our tumbler an all-black, all-white distribution. Now let us place a cover over the tumbler and shake it thoroughly. The result is a gray state. The interesting question now arises. Is it possible to shake this tumbler long enough to get all the black grains back again in the upper half and all the white ones below? And if so, how long a time is required?

Were we forced to attack this problem with the ordinary laws of mechanics, which are regarded as applying to the individual particles, an answer to this question would be impossible. The number of entities is so large that we could not follow their course even if we knew what their actual relations are. To appreciate this point, one has but to recall that we have no mathematical means of solving a simple mechanical problem for three bodies. Nevertheless a correct answer to our question can be given. The theory of probability makes it possible.

The brilliant French mathematician of the last century, Henri Poincaré has put the matter in the following paradoxical form. "You ask me to predict the phenomenon that will be produced. If I had the misfortune to know the laws of the phenomena, I would not succeed except by inextricable calculations, and I should have to give up the attempt to answer you; but since I am fortunate enough to be ignorant of them, I will give you an answer at once. And, what is more extraordinary still, my answer will be right." [19]

Poincaré's statement is correct. It is precisely because knowledge of the laws is impossible that an answer can be given. The largeness of the numbers which is one source of our ignorance, becomes an asset rather than a liability. For it makes it reasonable to suppose that one permutation or combination of our grains of powder is no more probable than any other. Wherever this can be assumed the theory of probability applies and its necessary consequences follow. The first notion which we must grasp therefore, is the principle that one permutation is as probable as any other. Upon this doctrine the theory of probability rests and only when it is valid does probability theory apply.

The reader may ask what justification science has for accepting this principle. Certainly it cannot be verified by direct observation. There are two reasons. One finds its basis in the first principles to which science has been driven. If the kinetic atomic theory is true it follows quite naturally from the large number and similar properties of the elemental particles, that, at least, when local details are neglected, one permutation should be as probable as any other. Second, when this is assumed we find that observed fact corresponds with its predictions. In fact, the degree to which this principle of randomness exhibits itself in diverse types of natural phenomena is truly remarkable. The hands of shuffled cards, the arrangement of matches relatively to cracks in a floor upon which they are thrown at random, and the chemical and physical properties of gases are a few cases in point. There can be no doubt that there is a principle of randomness or contingent relatedness at the basis of things. The empirical validity of the theory of probability and the range of application of statistical types of scientific law give unequivocal testimony upon this point.

This fact is of philosophical importance on two grounds. It gives the lie both to a thoroughgoing theory of internal or necessary relatedness, and to the complete adequacy of the operational theory [21] of scientific concepts. The first consequence is obvious. The second becomes evident the moment one notes that the scientific principle to the effect that one permutation is as probable as any other, is never defined by an operation; only certain of its consequences are amenable to such a type of definition. This is true of all the methodoligical philosophies of science such as pragmatism, instrumentalism, positivism, etc., which attempt to restrict science to a particular type of methodological procedure. There is no one method in science, and it would be a thoroughly unscientific state of affairs if there were. Sheer dogma would be the result. It is precisely as unscientific to fit facts to methods, as to theories. Only observed nature itself and the first principles to which it drives one can determine the methods which science can use. Some subject-matters entail one, some another. Only the materials in question can provide a basis for decision.

The theory of probability has one other consequence. At first sight it appears to be paradoxical. Its principle that one permutation is as probable as any other, necessitates that some events must be more probable than other events. It is this corollary which makes probability theory a fertile means of prediction, and enables us to answer our question concerning the grains of sand in our tumbler

To make this clear it will be well to reduce the number of grains. No error will arise from this since the principle is unaffected by the number of entities involved. All that we must be sure of is that our grains must be sufficiently alike intrinsically to insure that there is no more reason for one permutation of them to arise than another.

To simplify exposition we shall assume that our grains

of powder are twenty perfectly round balls, ten of which are white, the other ten, black.[18] We shall assume also that they are enclosed in a bulbous glass container from which a finger-like hollow glass tube projects, and that this tube is long enough to contain all of the twenty balls, one on top of the other. Also let each ball have a number so that the black balls number from one to ten, the white, from eleven to twenty. Suppose also that all are in their natural order from one to twenty with the black balls on the top, and the white on the bottom. This corresponds to the original state of our gray powder except that we are able, because each has a unique number, to distinguish between individual permutations of the balls or grains.

Now let us turn the container upside down, shake it thoroughly, and then allow the balls to run back again into the finger. What is the chance of getting all the balls back again in their original unique order? Our principle that one permutation is no more probable than any other enables us to give a very definite answer. It is one in the number of different ways in which twenty objects can be combined. The latter number happens to be very large. It is 2,432,902,008,176,640,000.[18] It will hardly be worth our while to try to regain our original permutation immediately.

But suppose that you do not insist on the original order, and will be satisfied if we get all the black balls in the upper half and all the whites in the lower portion. Then our chances are much better. For while there is only one way in which the twenty balls can combine to produce our original permutation, there are a large number of different ways in which they may arrange themselves to insure in all-black, all-white distribution. In fact the number is very large. It is 13,168,189,440,000. Thus whereas the original permutation represents a class with only one member, the all-black, all-white distribution represents a class with 13,168,189,440,000 members.[18] Since any one of this larger class is as likely to occur as our original permutation, it follows from the principle that one permutation is as probable as any other that the event which is any all-black, all-white permutation is 13,168,-189,440,000 times more probable than the event which is a specific one of this class, such as the specific permutation with which we started. Hence, to understand why the equal probability of permutations necessitates that certain events must be more probable than others, one has but to draw a distinction between a specific permutation, and events which are classes of permutations. Permutations have equal probability, classes of permutations can vary in probability with the number of members which they contain. This permits us to re-define the probability of an event in a new way. That event is the most probable which represents the class of permutations which has the largest number of members.

It is for this reason that most of the permutations of our ten balls will be grey. For while there are a large number of all-black, all-white permutations there are a greater number of gray ones. In fact, for any all-white, all-black permutation there are 184,756 different gray permutations.[18] Hence the probability of a gray distribution is 184,756 times greater than that of an all-black, all-white one. It is for this reason that we shall get gray most of the time if we shake either our twenty balls, or the tumbler full of powder.

The importance of the distinction between a permutation and a class of permutations, for an understanding of the theory of probability becomes clear if one centers attention on a specific gray permutation. The chance of getting it back, after a shaking of the twenty balls, is not one whit better than that of a specific all-black, all-white one. It is only to events representing classes of permutations which contain a different number of members, that differences in probability apply.

Consider these ideas in their bearing upon the nature of scientific law. To say that a law has a statistical interpretation is to maintain that the principle which the law expresses represents a class of permutation for the subject-matter in question, which is so large in number, compared to any other classes of permutations which are possible, as to be practically the general rule. But the very principle upon which the statistical interpretation rests insures that this general rule has its exceptions, and that given time enough they must occur.

Note what follows, therefore, from the current accepted statistical interpretation of the second law. To say that nature is running down means merely that a shuffling process is going on which will eventually insure, as G. N. Lewis has said,[15] that nature will eventually become perfectly shuffled. To use our terminology, nature is proceeding from the present state of organization which represents a class of permutations with more members than classes representing previous states of nature, to that class with the greatest number of permutations, which represents the eventual state of frigid chaos and dust.

Now let us imagine that nature has reached this state, dismissing all the difficulties concerning why we are not there already,—difficulties which are as fatal to the second law in this as in its literal interpretation. The most probable conclusion concerning the size of our universe indicates that the number of basic entities which it contains must be finite in number. The theory of the finite universe necessitates this. It follows, therefore, that the number of permutations which are possible must also be finite in number. Hence, since the motion of the microscopic entities is endless, nature has an endless succession of shuffles in which to exhaust a finite group of possible permutations. Therefore, after nature has run down to the state of maximum entropy, the original most improbable state of minimum entropy must again arise, since it represents a possible permutation which has actually existed. In other words, it follows as a necessary consequence of the accepted meaning of the second law of thermo-dynamics that the running-down must be balanced by a corresponding running-up and that this law must be true half of the time and false the other half.

We find, therefore, that a careful analysis of the real meaning of the second law reveals precisely that conclusion concerning its applicability to the whole of nature, which the macroscopic atomic theory necessitates. It is but one aspect of a two-fold process.

Moreover, the secondary and derivative character of time, which relativity physics has revealed makes it more reasonable to suppose that the breaking-down and building-up processes are in operation continuously, as Millikan has suggested,[20] than that the one follows the other in temporal sequence, as our traditional temporal form of expression has indicated. Nor is this conclusion without its factual justification. For its acceptance permits us to take the present state of organization and the process of organic evolution at their face value, without the introduction of an endless series of unverifiable ad hoc hypotheses.

Furthermore, we have but to accept the principle of polarity of the macroscopic atomic theory, which has already revealed itself in relativity theory and quantum and wave mechanics, to understand why both the organizing aspect and the disorganizing aspect to which the second law refers, exist together on a single planet in nature at the same time. The disorganizing effect and the statistical character of the second law is precisely what must arise from the contingent relatedness of the microscopic particles, whereas the maintenance of organization centers in the changeless form of the macroscopic atom. The microscopic factor constitutes the verified basis of the second law and makes chemical transformations and processes of disorganization possible; the macroscopic factor builds up systems to maintain the total organization of nature as a whole at a constant, and thereby insures that there are always organized systems present to be disorganized.

Thus a Creator or an extra dimension of space to provide organized materials for the operation of the disorganizing principle expressed in the second law is unnecessary. Also, it is because the organizing principle in living things has its ultimate basis in the macroscopic that the source of their energy is in the sun and stellar nature.

It appears that the error which has misled us in thermo-dynamics is the same one which misguided us in pre-relativity mechanics. We have taken the concept of time in nature as a whole too seriously. It is not an absolute principle. The second law of thermo-dynamics is a physical and logical principle, and not a temporal process. Nature is not a process in time headed to destruction, but an eternal opposition of two principles, one of which Sadi Carnot discovered when he formulated the second law, the other which we have come upon in our discovery of the macroscopic atom.

The significance of the macroscopic atomic theory is not exhausted with the new interpretation which it necessitates for the second law. It also brings theory into alignment with fact, in one other respect. As we indicated previously the notion of potential energy had somewhat the same status in thermo-dynamical theory that the principle of the equivalence of inertial and gravitational mass possessed in Newtonian mechanics. Fact called for the concept, yet the traditional conception of first principles provided it with no theoretical justification. Only the kinetic principle had a basic status. Our theory removes this inadequacy. Energy due to motion has its basis in the moving microscopic particles and energy of organization and position has its basis in the macroscopic atom. Again the polar principle which our theory provides reveals itself in technical science.

However, the most significant consequence of our theory appears in biology. If nature is a mixture of an internal compounding of microscopic atoms and a form imposed on the whole and through the equilibrium of the whole upon all the local parts it does not surprise us that living things are physico-chemical, and at the same time exhibit an organization which cannot be defined completely in terms of their constituent microscopic physico-chemical materials. Thus for the first time in Western science a solution of the age-old biological problem of organization is possible without accepting the matter-form or functional theory of nature of Aristotle and the denial of the primacy of the physico-chemical which it entails. Moreover, since there is nothing to prevent the microscopic tendency from getting into the ascendancy in one local process such as the inorganic, and the macroscopic organizing tendency from exhibiting itself in another process in the same region, the presence of organic evolution in an inorganic environment in which the second law operates does not present a problem. Thus we are able to get life into existence without the invocation of extra-physical entities and the postulation of metaphysical principles which are incompatible, or difficult to reconcile, with those that the more certain facts of inorganic science have necessitated.

Certainly the first principles of one must be those of the other. Living organisms and inorganic systems are part of the same universe. The first principles of this universe are the conditions of both. To suppose that the philosophy of one is different from the philosophy of the other is as ridiculous as to suppose that one can build a single building upon two entirely different types of foundation at the same time, or that one can have the delicate chassis of a Rolls Royce and the gross frame of a Mack truck in the same motor vehicle.

But we must let the organism speak for itself. Theory must be fitted to fact, not fact to theory. It may very well be the case that characteristics exhibit themselves in living things which entail additions or amendments to the first principles which inorganic science necessitates. It happens, however, that this is not the case. For the macroscopic atomic theory to which we have been driven by our analysis of relativity physics and quantum and wave mechanics, was discovered by the writer first in biology. It was only because it could not be true for biology unless it is also true for physics that he was led to the physical investigations which have produced the separate grounds given in this and the preceding chapter. The independent evidence from biology together with an analysis of the exact nature of a specific living process will appear in the next chapter.

References and Bibliography

1. J. Merz. History of Scientific Thought in Nineteenth Century. Blackwood.
2. J. Perrin. Atoms. Constable.
3. F. Reiche. The Quantum Theory. Methuen.
4. N. Bohr. The Theory of Spectra and Atomic Constitution. Cambridge Press.
5. De Broglie and Brillouin. Selected Papers on Wave Mechanics. Blackie & Son.
6. E. Schrödinger. Collected Papers on Wave Mechanics. Blackie & Son.
7. G. P. Thomson. Waves and Particles. Science. LXX. No. 1823.
8. W. Heisenberg. The Physical Principles of Quantum Theory. Univ. of Chicago Press.
9. P. A. M. Dirac. The Principles of Quantum Mechanics. Cambridge Press.
10. The Dialogues of Plato. Trans. by Jowett. Oxford Press.
11. Hegel. Science of Logic. 2 vols. Trans. by Johnston & Struthers. George Allen & Unwin.
12. M. Cohen. Concepts and Twilight Zones. Journ. of Phil. XXIV. No. 25.
13. W. H. Sheldon. Strife of Systems and Productive Duality. Harvard Press.
14. A. Einstein. Science 71:608–10.
15. G. N. Lewis. The Anatomy of Science. Yale Press.
16. Lewis and Randall. Thermodynamics. McGraw-Hill.
17. J. Jeans. Astronomy and Cosmogony. Cambridge Press.
18. C. E. Guye. Physico-Chemical Evolution, 30ff. Methuen.
19. H. Poincaré. Science and Method. P. 66. Thomas Nelson & Sons.
20. R. A. Millikan. Available Energy. Science. LXVIII. No. 1761.
21. P. W. Bridgman. The Logic of Modern Physics. Macmillan.
22. A. N. Whitehead. Science and The Modern World. Ch. VIII. Macmillan.
23. A. S. Eddington. Stars and Atoms. Oxford Press.
24. F. S. C. Northrop. Relation Between Time and Eternity, etc. Proc. of 7th Int., Cong. of Phil. Oxford, 1930.
25. G. N. Lewis. The Symmetry of Time in Physics. Science. 71: 569.

CHAPTER IV

The Living Organism

The living organism is an elusive creature. Even science has found this to be the case. Attempts at an adequate philosophy of the organism have never been fully successful. The problem of organization, which is the crux of the difficulty, has not had an adequate solution since Darwin gave the final death-blow to the Aristotelian philosophy.

ARISTOTLE, LAVOISIER, AND BERNARD

The two facts which constitute this problem are obvious. They were indicated by Hippocrates of Cos[1] in Greek times. He noted that in some fundamental sense, a living organism is a mechanical system. Diseases have discoverable and dependable causes, and understandable and predictable effects. This discovery made scientific medicine possible. Those who are prone to attack the mechanical emphasis of science as destructive of human values, will do well to reflect upon this point. Our capacity to alleviate human suffering would be in a sorry state were the principle of mechanical causation not applicable to life. The second fact which Hippocrates noted is organization. The relations which join the parts of a living thing together to form the whole seem to involve something more than the parts themselves. Since the presence of form which is not defined in terms of matter, is incompatible with the physical theory that is demanded by mechanical principles, the presence of mechanical causation and organization in living things, constitutes a problem. It is known as the problem of organization.

Empodocles' contributions [2] to it were noted in the first chapter. The first thorough-going attempt at its solution was made by Aristotle.[3] He approached living things from the descriptive point of view. The many kinds of organisms made classification necessary. Moreover, the form rather than the material provided the key to classification. It is easy to understand, therefore, why Aristotle concluded that life involves a formal as well as a material cause. In addition, he was greatly impressed by the fact of generation. Living things grow and reproduce themselves. There is hardly a word which appears more often in his inductive scientific treatises. We have noted how the immature state of physics and chemistry in his day left him no alternative but to take these two facts of generation and organization at their face value.

The result was a new philosophy of science, and a theory of the organism which dominated biological thought from the third century before Christ to the middle of the last century when Darwin [4] destroyed the doctrine of the fixity of types. The essential principles of Aristotle's philosophy follow necessarily from the two facts which he noted. To take generation as irreducible is to reject the principle of being of the physical and mathematical theories of Leucippos and Plato, for the principle of becoming. Reality is something which changes its properties, rather than an indestructible motion of atoms or an eternal system of mathematical relations. From this the principle of teleology follows, as we have noted. Likewise, to regard organization as irreducible, is to admit both form and matter as causes. But this is nonsensical unless both are regarded as mere passive attributes of the dynamic unity or efficient cause which renders their interaction intelligible. The action of a disembodied form on a separate embodied substance is meaningless. Only if the material and formal causes are attributes or expressions of the activity of the efficient cause, which, as we have noted, is teleologically controlled by the final cause, does the doctrine of a form which is irreducible to matter have meaning.

This is Aristotle's lasting contribution to biological philosophy. He saw quite clearly that the doctrine of real generation or emergence, and of organization which is irreducible to physical principles is incompatible with the mechanical and physical theory of Leucippos and Empodocles. Stated in modern terms this means that one cannot combine an irreducible emergence, or a non-physical organic relation, with the physico-chemical and mechanical conception of the living thing. Physiological chemistry and a non-physical organic philosophy, or emergent evolution and physico-chemical interpretations and mechanical methods are contradictions in terms. It cannot be too strongly emphasized that all emergence and relatedness must be defined completely in physical terms or the entire physico-chemical and mechanical conception of life must be overthrown.

Needless to say, the history of biological science since Aristotle gives no evidence that the latter alternative is justified. Whatever else may be true we must admit that a living organism is a physico-chemical system. With Harvey [5] the application of mechanical principles to gross physiological processes became evident. And with Lavoisier [6] their chemical and thermo-dynamical character was revealed.

The ideas of this young Frenchman will repay examination. He is one of the greatest scientists of all time. A little attention to his conceptions will save a vast deal of error, and many painful returns from long blind alleys later on. We can find the key to our problem if we will listen to what he has to teach us.

Coming to the living thing before specialization and analysis had called man's attention away from its general characteristics, he saw it as a whole in its relation to the rest of nature. With ordinary inorganic objects this is not important, but with life it is essential. For a living thing cannot exist in isolation. Its relations to the rest of nature are essential, not merely to its character and behaviour but to its very stability and existence. This is true of everything on the atomic level, as we have noted in connection with quantum and wave mechanics, but with living organisms it appears on the molecular and molar level as well. Lavoisier saw this. He perceived the living thing in its larger context. And what is more remarkable still, before chemistry or thermo-dynamics, as we know them, existed, he outlined the chemical,[20] thermo-dynamical,[21] and astronomical aspects of that context with an accuracy so complete that contemporary knowledge can add but minor improvements.

Sensing, with the insight of genius, the essential rôle which oxygen plays in most chemical changes, he developed a sound theory of combustion, thereby discovering the law of the conservation of mass, and placing chemistry upon those secure foundations which the following of chemical transformations with the balance makes possible. Perceiving the connection between heat and oxidation and between metabolic activity and oxygen supply he discovered that oxidation provides the key to metabolism and the release of energy in our bodies. Thereby the dynamical characteristics of animals became definitely connected with chemical substances and their interaction. Experimenting with animals in a block of ice which served as a crude, but nevertheless adequate, calorimeter, he and Laplace, established a connection between physiological changes of different kinds and the amount of heat given out. Thus the basis was laid for bio-physics and the thermo-dynamical conception of vital processes.[21]

But the oxidation of food and the release of its stored energy leads out to the synthesis of the food and the storage of that energy. The observation that animals depend either directly or indirectly upon plants for their livelihood guided Lavoisier to the botanical side of the biological world, and to the discovery of the reciprocal relation existing between plants and animals, in which the former breathe in carbon dioxide which they synthesize by the aid of energy from the sun with materials drawn from the soil to produce carbo-hydrates, giving off oxygen in the process; while the animals breathe in oxygen to oxidize the carbo-hydrates taken in as food, thereby releasing energy derived from the sun, and giving off carbon dioxide in return. Finally the materials taken up by plants and animals are "restored to the mineral kingdom by the breaking-down processes of fermentation, putrefaction and decay." Certainly the inorganic basis of life was clearly grasped by Lavoisier. It is no wonder that he founded experimental stations for the advancement of agricultural chemistry.

Consider the theoretical significance of these conceptions. Not only do they reveal the physico-chemical character of the vital, but they designate those respects in which a living thing differs from its inorganic neighbors. A stone or this book can be placed in a vacuum with no noticeable effect upon its stability or observable properties; but not so with man. His body is not sufficient unto itself. The sources of its energy and its temporary stability are external as well as internal. Ordinary observation misleads us upon this point. The apparently solid relatively permanent character of our bodies leads us to regard their stability as similar to that of the stone. We have but to isolate ourselves from our atmosphere for but a few minutes to demonstrate however that this is not the case. The invisible oxygen molecules of our environment are absolutely essential to our existence. Only if life is viewed from the physico-chemical point of view can its most elemental characteristics be understood. An organism without the invisible gases of its atmosphere is more helpless than a street-car without its trolley, for the street-car merely loses its capacity to function, whereas the organism lets go of its very existence under such circumstances. We may summarize this point, first clearly grasped by Lavoisier, by saying that a living thing is a temporary dynamic equilibrium between chemical materials external and internal to its epidermis. To understand life without an appreciation of the chemistry of gases is impossible.

It is useless to attempt to understand living things until this point is fully grasped. They are not self-sufficient members of an aggregate, subject merely to stimuli from and adjustments to their environment. Even their own stability has one pole of its foundation in the environment.

There is no error greater than the common supposition that the physico-chemical theory of life necessitates that one conceive of it as a machine. No analogy is further from the truth. It is because of a failure to understand even the most elementary conception of a living thing which Lavoisier outlined, that most of the modern "disproofs" of the physical theory of life are attacks upon straw men. Writers use the deceptive character of ambiguous words to cause their readers to believe that the mechanical and physico-chemical theory of life involves the conception of living things as a rigid machine, and then imagine that they have established a proof for vitalism or teleology when they have indicated that such a crude analogy is inadequate. Had they mastered what Lavoisier discovered at the end of the eighteenth century they would have been saved all this trouble, and have become aware that the real weakness of the physical theory of life is in their own failure to grasp what it is.

This brings us to the heart of the difficulty concerning the first principles involved in living things. Man has been attempting to develop a theory of a living organism before he has determined its essential general characteristics. Since the time of Lavoisier, it has become increasingly evident that a living thing is a physico-chemical system involving a temporary dynamic equilibrium between internal and external physico-chemical materials. The real task is to determine what the general characteristics of this equilibrium are, and to trace them to their foundations.

Following Lavoisier, intensive analysis of specific organic compounds absorbed the interest of physiological chemists. Claude Bernard[7] called their attention back from an analysis of organic substances to the organization of the living thing as a whole, to reveal that while we know that life is to be approached by experimental methods and physical and chemical analysis, we are by no means possessed of an adequate theory of the organization of these materials. Certainly it places slightly too much of a strain upon one's credulity to believe that the organization of life is a mere accidental by-product of the motion of nothing but the microscopic atomic particles. Life is physico-chemical in character yet the problem of organization remains unsolved. Hans Driesch[8] made this clear in experimental embryology.

The dictates of sound procedure seem to be obvious. We must determine precisely what kind of physico-chemical organization we find in living things. With this information before us it is then a mere problem of formal logic to determine whether our traditional theory of the first principles of inorganic science is sufficient to produce such a system.

At first thought, the possibility of determining what the essential entities and relations of a living organism are, seems to be out of the question. For physico-chemical analysis has not gone far enough to give us all the chemical materials of a living thing. Moreover, the number of different materials in a whole organism is so large and their relations are so complex that we would have difficulty in grasping the major outlines even if we possessed the information. The difficulty seems to be insurmountable.

Upon further reflection, a ray of hope appears. We are interested only in first principles. Whether one kind of chemical material or another is involved in life does not concern us. As long as it is chemical, the theory of the organism is not affected. Moreover, it is only the general differentiating characteristics of biological organization that are essential to our problem. The precise details concerning how it arose, or how it was made precisely what it is in a concrete instance, can be left for later technical investigation to determine. All that we need to know now, is what the general outlines of biological organization are, and whether the traditional first principles of physical and chemical science have the fertility to produce such organization. In other words, we can attack the problem of causation from the logical rather than the temporal point of view.

It happens to be the case, as far as first principles are concerned, that a constituent process of a living thing stands in much the same relation to the organism as a whole, as the organism itself stands to its immediate inorganic environment. Furthermore, the same facts of organization and apparent teleological activity which characterize the organism as a whole and constitute the real difficulty for any philosophy of life,—these same teleological and organic attributes apply to constituent organic processes. Hence, as far as the theory of first principles is concerned, a specific organic process within an organism is as good as the organism as a whole.

It happens that we have fairly accurate information concerning the general nature of the organization of a specific physiological process. It happens also that the American physiologist, L. J. Henderson, has succeeded not only in determining the major constituents and relations of such a process, but also in exhibiting the many physico-chemical constituents and relations of the system as a whole, in their actual functioning organic unity and interrelationship.[9] Moreover, this exhibition is expressed in objective geometrical terms, which are correct in a quantitative as well as a qualitative sense, so that the precise nature of an organic system is made explicit, and the chance of corrupting our conception of it, by subjective judgment and uncontrolled imaginative construction is reduced to a minimum. For the first time in history Henderson's nomogram of the blood enables us to get a specific functioning living organic process before us in explicit, detailed, unified terms. An analysis of components has not been gained at the expense of unity. It is no longer necessary to take one bit of information concerning life from here, and another bit from there and run the risk of begging the question concerning philosophical implications, in the manner in which one fits the details together. We can begin with an objectively presented graph of a biological system, with its major chemical constituents specified and organized as nature has actually compounded them, and then adjust our theory of first principles to our findings. Such is the significance of Henderson's nomogram of the blood for the philosophy of the organism.

A few remarks concerning the general characteristics of the blood system are in order. Reasons for regarding it as a representative type of protoplasmic system have been given by L. J. Henderson.[9] We shall limit ourselves instead to fairly general characteristics. The general physiology of the circulatory system is well known. The function which it serves is also quite simple. Oxygen is taken up by the blood in the lungs and carried through the arteries to the tissues where it is dropped, and carbon dioxide is picked up to be thrown out upon arrival at the lungs.

The manner in which this ties in with the animal-vegetable interchange, that Lavoisier described, is clear. Without oxygen reaching the tissues to oxidize the carbohydrates taken in as food, metabolism would be impossible and life would be out of the question. Moreover, unless the carbon dioxide which results from this process of oxidation is removed, death ensues. The teleological character of the blood system is as marked as the apparently purposeful behaviour of the organism as a whole. Like a kindly baggageman on a moving train, the blood picks up oxygen where it is available and drops it where it is needed, and takes on carbon dioxide when its accumulation would be injurious to life, to throw it out later at the lungs where it is not injurious. Could anything be more teleological?

Moreover, a very delicately balanced adjustment between all the chemical constituents which are involved, is preserved throughout the entire transaction. This organic adjustment is expressed in technical language by the statement that the hydrogen ion concentration, not merely of the blood but of the body as a whole, is preserved at practically a constant. In fact, the teleological tendency of life may be defined in physico-chemical terms by saying that the body tends to maintain its hydrogen ion concentration at practically a constant. Let any marked deviation occur and life is no more.

The importance of this point has been expressed most eloquently by the English physiologist J. S. Haldane. His statement also serves to bring out the difficulty for traditional first principles, which biological systems present. Speaking of a certain bodily system he writes: It is "a "mechanism" tuned to react with astounding delicacy and constancy to the minutest changes in the alkalinity of the blood. This supposed mechanism consists of what is called 'protoplasm'; and protoplasm is something which from the physical and chemical standpoint is excessively unstable. . . . Yet this unstable mechanism reacts in the human body, hour after hour, day after day, year after year, true as tempered steel, to one absolutely definite and absolutely puny stimulus!" [13] A constancy is present which has its basis, not in the materials, for protoplasm is unstable, but in the organization of those materials. This is but another way of stating what we pointed out in connection with Lavoisier. A living thing is not a member of an aggregate able to exist by itself, but it owes its own existence to its relation to materials external to its own enclosing epidermis. In short, the sources of organic stability are not solely internal. We find therefore that the relation of the organism to its environment is analogous to the relation of the chemical atom to its field of radiation. Organism and environment are factors in a single system. This organic relationship between organism and environment,[14] and between a specific bodily process and the organism as a whole entails an important consequence. The functioning of a given process seems to involve all other processes. Furthermore, the consequences of the action of certain chemical materials seem to depend upon their relation to all other materials. We can understand, therefore, why Haldane and others have maintained that it is relatedness, an organic and distinctly biological principle, and not chemistry that is fundamental in life. Two contentions have been made by him in this connection: First, prediction is impossible on mechanical and physico-chemical principles, and second, all physico-chemical relations, and even the properties of chemical materials, are relative.[15] To seek for physico-chemical explanations is to be led into an endless maze. One factor depends on a second, the second on a third, and so on without limit, he says. We must understand, therefore, that any biological system such as the blood is very complicated. We can appreciate also why J. S. Haldane concludes that "the attempt to analyze living organisms into physical and chemical mechanism is probably the most colossal failure in the whole history of modern science." [13] Reasons for rejecting this conclusion will appear, but the considerations which caused it to be uttered are sound: One constituent of an organic system is intimately connected with any other, and predication is impossible unless this connection is understood and taken into consideration.

In selecting the blood for study as a representative organic process we can be sure that it presents the major physiological characteristics and philosophical difficulties of the organism as a whole. It exhibits teleological characteristics to a marked degree, it is a very complicated organization, the sources of its physico-chemical equilibrium are not purely internal, and it maintains a very delicate balance between its many constituents during many variations in the external and internal relations of its components. In addition it contains haemoglobin which is one of the two most complicated organic compounds that we know. If anything, the problem of biological stability and organization is more difficult than it is in the case of the more solid chemical substances which we usually have in mind when we think of specific organs and the body as a whole. What kind of a physico-chemical organization do we find this organic system to be?

L. J. HENDERSON'S NOMOGRAM OF THE BLOOD

The reactions of physiologists to this nomogram have been most amusing. Some have criticized it because it is true. It is but an unusual way of stating something that any well-trained physiologist knows, they say. Certainly, this is all to the good. One may add that the nomogram has the advantage of getting these ideas out of the imaginations of scientists onto paper in objective form, where we can know precisely what they are. Others have been unable to understand it. There is perhaps no better sign than this of the immature stage of development of biological as compared to physical science. The desirability of stating natural processes in geometrical form is taken for granted in physics. Certainly, the difficulty of understanding the nomogram is not due to the complexity of its mathematics, for it involves no ideas more advanced than those that an ordinary school-boy learns in his first lessons in analytical geometry. In fact, the manner in which the nomogram is built up is a fascinating subject. Furthermore, it is not without its relevance for an understanding of the nature of biological organization.

Let us begin, however, with the organism itself, centering attention upon the blood in its relation to respiration, metabolism, and the removal of waste products. Physiological chemistry has revealed a series of discoveries. If we grasp each one as we proceed, the nomogram will take care of itself, and appear as a welcome adjunct at a later stage of our analysis.

The first theory of the chemistry of the blood in its relation to oxygen intake was very simple. A complex organic compound in the red corpuscles, called haemoglobin, was supposed to combine with oxygen according to the formula

$$Hb + O_2 = HbO_2.$$

There can be no doubt that haemoglobin does combine with oxygen. In fact, the per cent of oxygen which is taken up with different degrees of oxygen pressure is known. It appears in the following graph.

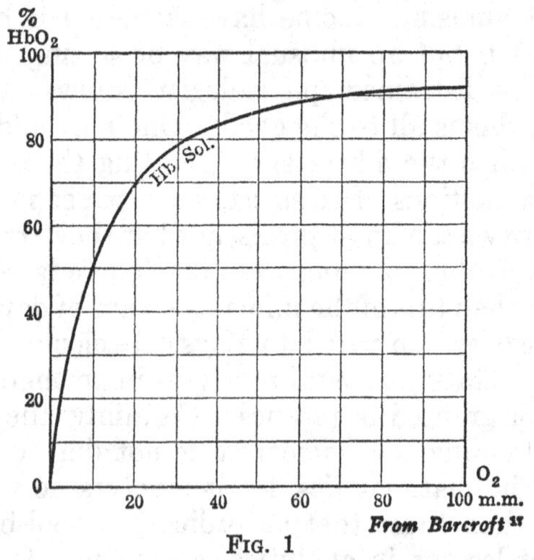

FIG. 1

This graph [17] is not without its philosophical significance. It is known that oxygen affects the hydrogen ion concentration. One would expect therefore, that great variations in oxygen tension should upset this concentration and cause death. The fact that the organism preserves itself through great increases of oxygen pressure seemed, therefore, to give the lie to the chemical theory of respiration, and suggested that a teleological principle is in operation. But the curve for oxygen absorption throws new light on this problem. It will be noted that with an increase of oxygen tension from zero to thirty there is an increase in percentage of combination from zero to eighty and that with a tripling of oxygen pressure from thirty to ninety there is an increase of only one-eighth in the oxygen that combines. In fact, the graph indicates that an increase in oxygen pressure above sixty m.m. of Hg gives practically no increase whatever in the amount of oxygen that is taken up. It appears, therefore, that the capacity of the body to preserve its existence through relatively great variations in oxygen tension has a chemical rather than a teleological basis. It centers in the peculiar chemical properties of haemoglobin.

Notwithstanding the part which haemoglobin plays in the absorption of oxygen by the blood, it becomes evident that the formula given above is not a complete account of it. For Barcroft[17] found that the amount of oxygen which is absorbed is also a function of the amount of carbon dioxide that is present. The following graph, in which the curves, in order from left to right, represent increasing tensions of carbon dioxide, establishes this fact.

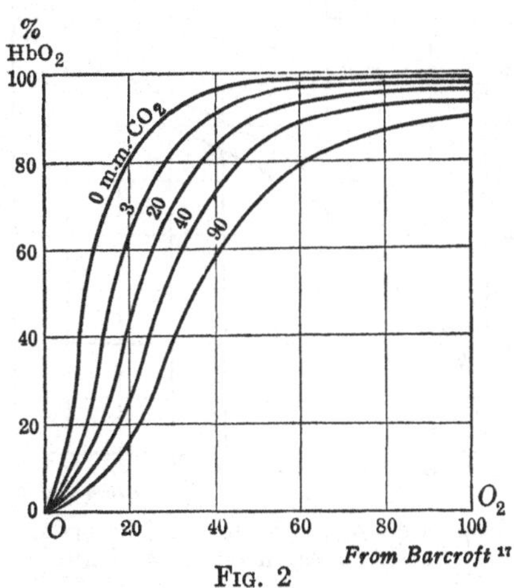

Fig. 2

It is to be noted that with an increase of carbon dioxide pressure from 0 to 90, the capacity of haemoglobin to hold oxygen in an atmosphere of 20 m.m. of oxygen pressure decreases from 80 to 20 per cent saturation. The philosophical implications of this point are interesting. It throws light on the capacity of the blood to behave teleologically like a baggageman. In the lungs, the tension of oxygen is high, while that of carbon dioxide is low. Hence the haemoglobin takes up a large amount of oxygen. In the tissues exactly the reverse is the case. When the blood reaches them it strikes low oxygen pressure and very high carbon dioxide pressure, as a result of the absorption of oxygen in metabolism and the production of carbon dioxide. Since high carbon dioxide pressure decreases the capacity of haemoglobin to hold oxygen, it follows on purely chemical grounds that the blood must release the oxygen in the tissues which it picked up in the lungs. Thus an apparent teleology is found to be conditioned by purely mechanical and physico-chemical factors.

Lest we forget that the amount of combined oxygen in the blood is a function of both oxygen supply and carbon

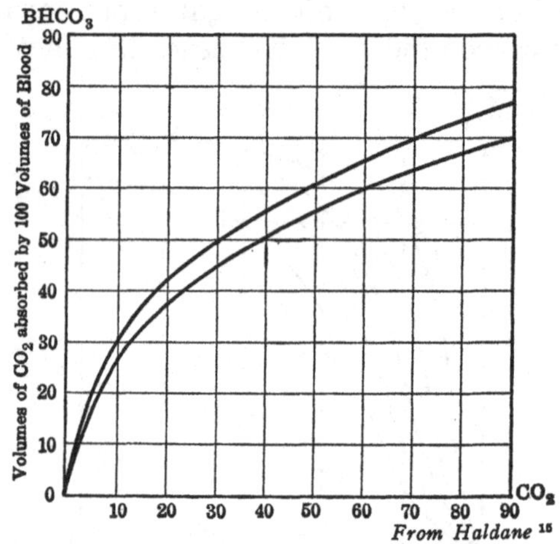

From Haldane [15]

Upper curve, in presence of hydrogen.
Lower curve, in presence of air.

FIG. 3

dioxide tension, let us express the triadic relation, which is involved, in the form of an equation.

$$f(HbO_2, O_2, CO_2) = 0$$

This means that when the values of any two of these three variables is known the third is specified by the equation.

But J. S. Haldane, Christiansen, and Douglas also found a similar relationship to hold between carbon dioxide, combined carbon dioxide or carbonate, and oxygen.[15] The graph in Fig 3 expresses this fact. The curves show that the capacity of the blood to take up carbon dioxide to form the carbonate is decreased in the presence of oxygen. Thus we find another chemical basis for the apparent teleological capacity of the blood to pick up and drop oxygen and carbon dioxide where such activity is most beneficial. In the tissues the carbon dioxide tension is high, the oxygen pressure relatively low, hence the blood will absorb carbon dioxide there. In the lungs, the reverse is the case. High oxygen pressure is present. Since this decreases the capacity of the blood to hold carbon dioxide, we can understand, therefore, why it is released at the lungs. Another triadic equation expresses this interrelation.

$$f(BHCO_3, CO_2, O_2) = 0$$

In this equation there are two chemical factors, one, the salt; the other carbon dioxide, which combines with water to form a weak acid, H_2CO_3. In solution such a weak acid tends to break up into the positive hydrogen ion and the negative alkali ion. The degree to which this dissociation occurs depends on the amount of the salt of this acid which is present. It follows, therefore, that the relation between carbon dioxide and the carbonate or combined carbon dioxide will determine the hydrogen ion concentration. Thus Henderson in his analysis of the blood system was led to a third triadic relation

$$f(CO_2, BHCO_3(\overset{+}{H})) = 0$$

where CO_2 equals carbon dioxide, $BHCO_3$ the carbonate or salt, and $(\overset{+}{H})$, the hydrogen ion concentration. This relationship can also be represented in a two dimensional graph.

At this point another consideration appeared. It was known that a variation in the tension of carbonic acid is accompanied by a variation in the passage of electrolytes between the corpuscles and the serum of the blood. Since we already know that carbon dioxide tension is a function of oxygen supply, it follows that the amount of electrolyte, such as a chloride, which is present, must be a function of oxygen as well as carbon dioxide tension. Thus a fourth triadic relation was discovered and graphed and verified. The following equation expresses it.

$$f(CO_2, O_2, BCl) = 0$$

where BCl represents a chloride.

In this fashion a physiological process which was first thought of as involving little more than the simple relation between haemoglobin and oxygen was found to involve four equilibria, each of which involves three chemical components. Note how this physiological process is becoming more and more complex as we examine it. The reader will undoubtedly appreciate Haldane's charge that the physico-chemical theory of life leads one into a maze of confusing relations.

However, Henderson did not stop at this point. Four triadic laws were present.

$$f(HbO_2, O_2, CO_2) = 0$$
$$f(CO_2, BHCO_3, O_2) = 0$$
$$f(CO_2, BHCO_3, (\overset{+}{H})) = 0$$
$$f(CO_2, O_2, BCl) = 0$$

He noted that while these four triadic laws involve twelve terms, only six of the twelve are unique. The reader will note that they are O_2, HbO_2, CO_2, $BHCO_3$,$(\overset{+}{H})$, and BCl. But six different factors combined by three's can combine in twenty different ways. It follows, therefore, Henderson reasoned, that we have but begun to discover the actual relations within this physiological process. There must be sixteen other triadic laws in addition to these. Experiment confirmed this prediction.[9]

We are thus led by his reflections and experimental

work to a new conception of the blood system as it functions in connection with the external environment and other physiological systems of the body: It is a twenty-sided equilibrium in which six major constituents are joined by twenty different triadic relations. Undoubtedly, this neglects certain minor factors which must be included. In fact, more recent investigations by Henderson have altered the constituents somewhat. But the major general conception of the nature of a biological system remains the same. As far as first principles are concerned, the account which we have given is representative and sound.

Once one becomes accustomed to this somewhat technical account of the nature of a living system, points of philosophical significance begin to appear. First, intricate organization is present to a marked degree. Second, the strictly physico-chemical character of apparently teleological processes is revealed. Third, the real reason appears for Haldane's charge, that physico-chemical reasoning breaks down. It is not that the physical theory of life does not apply, but that the conception, to which it leads one, involves more variables and relations than the mind can keep in attention at once. It is impossible for the average mind to keep six factors and twenty triadic laws between them in consciousness at the same time. But this is evidence that we must bring mathematical tools to the aid of our intellect; and no proof whatever that the physical philosophy of life has failed. We do not reject the physical theory of nature in the inorganic realm because relativity theory is somewhat complicated; there appears to be no reason for doing so in biology. Fourth, there seems to be no evidence that physico-chemical materials become fluid and relative in living things. Many of these laws are established in the laboratory with inorganic materials. They hold equally well for living systems. Finally, the charge that one cannot predict in biology if the physical theory is accepted seems to fall to the ground. Were this charge valid it would be hard to understand how sixteen laws could be deduced from the four which we have given, and found to hold.

There seems to be no alternative but to face the conclusions to which experimental analysis leads us. This one process of the living organism involves twenty triadic relations between six different variable physico-chemical components. But the blood system is not twenty relations; it is all of them in one relationship. To understand this living process we must grasp these facts as a whole. The mind can only take one aspect at a time. It follows, therefore, that our capacity to really grasp the essential nature of this physiological system depends upon our success in getting some graphical means of preserving these six variables and their twenty triadic relations before us as a single system at once. Henderson gave his attention to this problem. The result was his nomogram of the blood.

The fundamental ideas which constitute it are quite simple. He noted that each one of the twenty triadic laws can be expressed objectively in a two dimensional Cartesian graph. Such graphs for two of these laws were given in Figures 2 and 3. Consider Figure 2. The vertical axis represents different percentages of combined oxygen; the horizontal axis, different amounts of oxygen pressure, the different curves, varying amounts of carbon dioxide pressure. The important property of such a Cartesian graph is that a given point in it specifies the values of all three of these variable factors. But a point in a two dimensional graph is specified when two quantities are known. Hence the experimental specification of any two of the three variables in question, enables us to determine the value of the third without experiment, by the aid of the graph. Thus if we know that the carbon dioxide tension is 40 m.m. and the oxygen pressure is 20 m.m., the graph tells us that a 50% oxygen saturation exists.

But a determination of the values of any two of our six

unique variables not only gives us the value of a third, but also the values of the other three. An examination of the four laws with which we started will make this clear.

$$f(O_2, HbO_2, CO_2) = 0 \quad (1)$$

$$f(CO_2, BHCO_3, O_2) = 0 \quad (2)$$

$$f(BHCO_3, CO_2, (\overset{+}{H})) = 0 \quad (3)$$

$$f(CO_2, O_2, BCl) = 0 \quad (4)$$

We have noted that the graph for the first law enables us to determine the value of HbO_2, if we know the values of O_2 and CO_2. But with a knowledge of CO_2 and O_2, equation (2) gives us the values for $BHCO_3$. With $BHCO_3$ and CO_2 known, equation (3) gives us the value for $(\overset{+}{H})$, and with CO_2 and O_2 known, equation (4) gives the value of BCl. Thus the value of any two of our six variables gives us the values of the other four. Since a two dimensional Cartesian graph is capable of expressing all possible values of any two variables, and the values of two variables enable one to determine the values of the other four, it follows, therefore, that the twenty triadic relations between these six variables can be expressed in terms of a two dimensional Cartesian graph. The French mathematician D'Ocagne has devised graphs of this character. They are called nomograms.

The manner in which a nomogram is constructed is quite simple. Let us begin with two variables and choose out of our six the carbonate $BHCO_3$ and the hydrogen ion concentration $(\overset{+}{H})$, designating the former by the vertical axis; the latter by the horizontal one. We have the graph in Fig. 4. Any specific value of these two variables automatically determines a point on this graph. But we have noted also that the values of two variables automatically determines the values of four other variables. Hence any point which these two variables specify must also determine the values of the four other variables. To do this graphically it must represent a value in a series of O_2, HbO_2, CO_2 and BCl lines, as well as in the series of

$BHCO_3$ and $(\overset{+}{H})$ lines given below. It follows therefore, that the graph below must be complicated by the addition of four other series of lines representing different possible

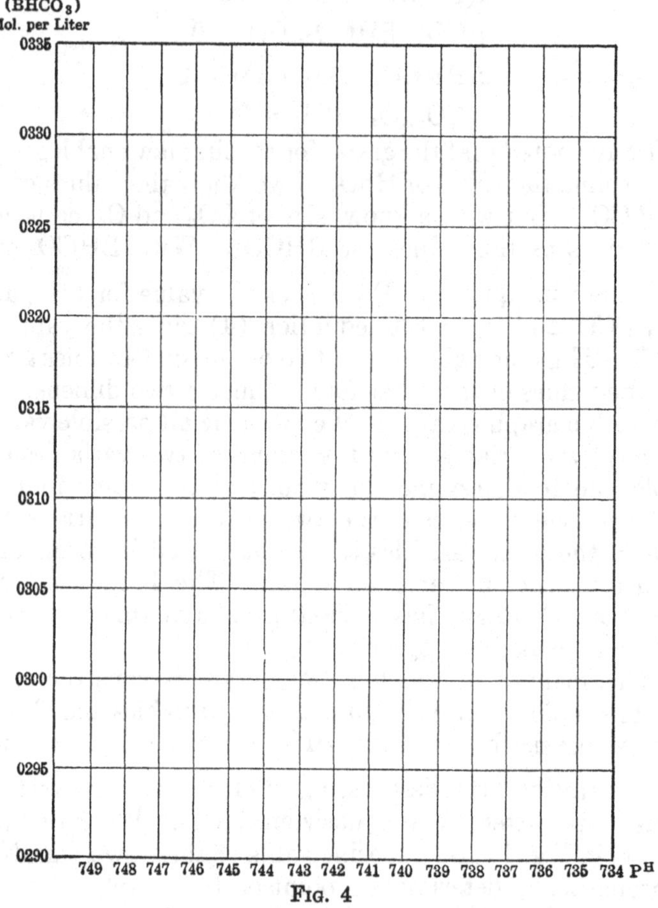

FIG. 4

values of the other four variables. Actual experimental evidence as expressed in the four separate graphs which were first discovered, designate what the precise lay and distribution of the four additional series is. Actually for the blood it is as follows:

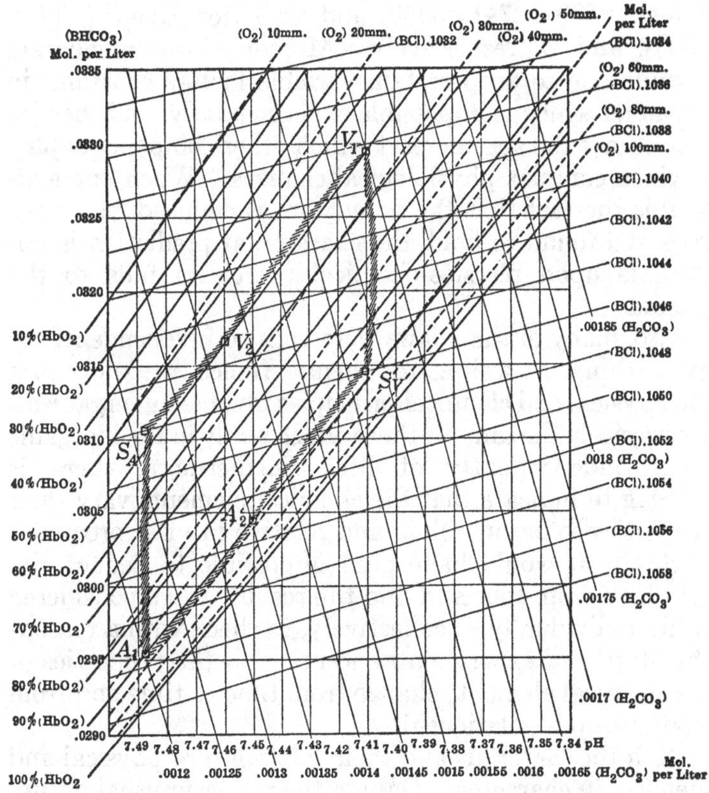

From Henderson,[9] with heavy broken lines, indicating the arterial-venous-serum relationship, added by F.S.C.N.

FIG. 5

The immediate apparent complexity of this nomogram need not confuse anyone. For the aid which it gives to the mind is far greater than any difficulties which it puts in the way. The important point to note is that when any two of the six variables contained in this graph are specified, a point is determined which automatically enables one to read the values of the other four. It is to be noted that the numbers attached to the lines of the graph represent experimentally determinable quantities. For example, an H_2CO_3 value of .0016 and a BCl value of .1036-determine the point marked V_1. This point then gives the values .0330–, .741, 50%, and 25% for $BHCO_3$, $(\overset{+}{H})$, HbO_2, and O_2 respectively. Moreover, the nomogram permits one to keep certain variable factors constant, in a manner which is impossible experimentally, and thereby determines the specific part which other components play in given complex physiological changes. When one adds to this the fact that the nomogram is deduced from four laws and found to hold, the charge that prediction is impossible upon physico-chemical principles falls to the ground.

Also, many of these laws were verified independently of living things as well as for them. Hence, we know that the physical and chemical principles involved in organisms are precisely the same as those which apply in the inorganic realm independently of life. Furthermore, there is nothing to indicate that the chemical elements vary their properties to become fluid and relative in vital processes. Certainly it would have been impossible to deduce sixteen laws from four and find the results, when considered either individually or collectively, to check with a reasonable degree of experimental accuracy, if the properties of the chemical elements change from time to time and from relationship to relationship.

Both the materials and their relations are physical and chemical in character. Only in the rather unusual nature of the complex system which they produce, does life exhibit its uniqueness. And even here nothing non-physical or organic seems to be involved. No additional biological organizing principle is necessary. For the whole compounds out of the parts by the same geometrical and chemical rules which apply in the inorganic. Otherwise, it would not have been possible to fit the graphs of the twenty different equilibria together to form the nomogram and attain a unified mathematical picture of life which holds with a reasonable degree of quantitative accuracy. Life in its organic unity and constituent relations and materials involves nothing but physical and chemical principles.

It by no means follows, however, that our philosophy of the organism is complete. In fact, the nomogram informs us that it is not; for a type of physico-chemical system is revealed which our traditional theory of first principles is incapable of explaining. This brings us to the consideration of the kind of physico-chemical system which the nomogram reveals.

THE NOMOGRAM AND THE NATURE OF A LIVING SYSTEM

Up to this point we have been considering what the materials and their interrelations are, and how they were built up in the nomogram to constitute the organic process as a whole. We now turn to an examination of the essential characteristics of this whole.

A consideration of the difference between a nomographic picture of a stone and of our organism will prepare us for the correct conclusion. Consider the nomogram again. It will be noted that a large dot marked V_1 appears in the upper portion of the graph, and that a similar dot marked A_1 appears in the extreme lower left-hand corner. These two dots specify the physico-chemical constitution of venous and arterial blood respectively. Since the blood in its circulation is undergoing a continuous cyclic change from arterial to venous and back again, it follows that the real nature of this system as a whole is to be represented by a continuous orbital line extending from the point A_1 to point V_1 by one route and back by another. The accompanying graph gives a rough representation of such an orbit.

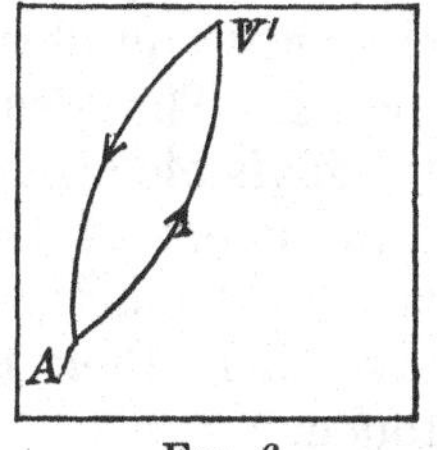

FIG. 6

A consideration of precisely what this line means will reveal that it represents continuous combinations and decompositions of chemical materials. We are thus led to the conception of a living organism as a cyclic process involving a flux of chemical combinations and disorganizations. The latter point is important. Chemical decomposition is as essential to the stability and organization of life as chemical attraction. This is of great philosophical importance. It reveals that life is a dynamic physico-chemical equilibrium and that the source of its stability is not to be found in purely internal bonds of physical or chemical attraction. Thus we come again upon the fact noted in connection with Lavoisier. The source of the stability of a living system is not purely internal. It is a dynamic equilibrium between internal and external physico-chemical elements.

The external aspect of this equilibrium appears explicitly in the nomogram. One of the most important of the six major components of this system is oxygen. Moreover, it is not merely to oxygen as oxygen that the nomogram refers but to its pressure. Now the pressure with which oxygen comes to the organism is not a function of anything within the epidermis of the living creature; it is conditioned instead by the astronomical and geological forces which have determined the constitution of the earth's atmosphere. As far as its essential characteristics as an actual living thing are concerned, the nomogram indicates unequivocally that an organism is determined as much by the oxygen of its environment as by any other factor within it. If anything, the rôle of oxygen is more important than that of the other variables, although all of them are necessary. This is what we mean when we say that a living thing is a dynamic physico-chemical equilibrium between external and internal elements. The idea which one simply must grasp, is that the invisible atmosphere which we breathe and the whole atmosphere of the earth which determines its tension is just as much a part of our bodies as the bones which give them their rigidity or the flesh which we see and touch.

The difference between our living bodies and this book or the table on which it rests, is not that one is a physico-chemical system whereas the other is not; but that nature, as it moves its matter and energy about, has achieved temporary stability in one in a manner somewhat different from that used in the other.

A nomographic representation of the physico-chemical character of this book or the desk will serve to make this clear. We have noted that a living thing involves a continuous series of points for its representation; a single point on a nomogram would suffice however to represent the stone. One would merely determine the amounts of its different constituents, designate the point of the graph which they define, and one's work would be done. Chemical disorganization is not essential to its character. On the chemical level, the sources of its stability are internal. Except for very slight exceptions it does not exhibit itself as a balance between external and internal factors until one gets down below the chemical level to the organization and structure of wave and quantum mechanics. To be sure this rigid distinction between living things and the ordinary physical objects of our immediate environments needs to be softened down somewhat, but for the most part it is essentially true. Unlike the stone, a living thing cannot be conceived as sufficient unto itself; the source of its stability is not due solely to chemical affinity and attraction. Were this the case metabolism would not exist and its dynamic characteristics would be impossible. Instead it is a temporary dynamic equilibrium between internal and external physico-chemical materials.

This fact is very important, because it means that we must look to the foundations of inorganic nature itself, to the ultimate principles which move matter and energy about, for the source of organic stability. Once one senses the real nature of a biological system it becomes as ridiculous to look solely to the microscopic chemical materials within the organism itself for the source of its stability, as it would be to hunt for the repose of the Arch de Triumphe in but one foundation of its support. Nature has put a bit of all of itself into even the most humble living creature.

This is not poetry, but concrete fact. One has but to begin with the nomogram, note the significant rôle of the pressure of oxygen, and trace that out to the geological and astronomical processes which condition it to discover that the astronomical principles of the whole of our universe are internal to life. This point has revealed itself in many other places. It resounds with eloquence when one considers energy and traces it back through food and plant to the sun and the source of its terrific energy supply. In Henderson's evidence suggesting "the fitness of the environment" [10] the same intimate connection between the character of life and the most general and universal properties of the inorganic reveal themselves. Here, more than anywhere else, the attempt to resolve the problem of the philosophy of the organism by insisting that biology is an autonomous science reveals its futility.

These considerations indicate that if the organization and apparent teleology present in living things involve something more than traditional principles, a purely biological energy or relation, or a psychological entity such as entelechy is wholly inadequate as the missing factor.[18] For the organization and adaptation is not local to life; it entails the most universal character of inorganic nature which forms one pole of its very existence. Life is not an after-thought produced by the mysterious appearance of a new principle, after the inorganic world was formed. Neither is it the handiwork of a disembodied soul or form perched in some mysterious fashion in the midst of chaotically moving particles. Nature is neither so poverty-stricken in her resources, nor so demoralized and ruffled in her spirit as to be driven to the undignified level of the irritated washerwoman impatiently shaking her finger at the playful children of the neighborhood, in order to get order into her community. The organization and stability of living things has its foundations too widely extended in the depths of inorganic nature to permit one to rest his biological philosophy on irreducible biological or psychological categories. The nomogram joins with other evidence to confirm this conclusion when it reveals that a living thing is a temporary dynamic heterogeneous equilibrium between internal and external physico-chemical materials.

The nearest analogy to it in ordinary experience is the whirlpool. The latter system does not owe its stability to an internal attraction between the drops of water which compose it. The drops within its form are but one aspect or pole of its temporary equilibrium; the other is to be found in the forces that move matter about in the vast ocean that surrounds it. The organism is the same, except that the external pole of its stability extends beyond any local region or local tidal movement to the basic principles which condition geological and astronomical systems. Moreover, the whirlpool is but a homogeneous physical equilibrium whereas the living organism is a heterogeneous physico-chemical one. Differences of chemical material, and chemical changes involving decomposition as well as composition, are essential to life.

The interesting question now arises. This equilibrium because of its physical nature and the part which chemical decomposition plays in it, is not held together by forces of internal attraction. What then is the source of its stability? Why during all the variations in food and temperature to which it is subject, and through the continuous chemical decomposition which goes on within, does it preserve its identity at all? How out of such unstable and active chemical materials has nature ever succeeded in producing a system that will preserve itself?

We can put these questions in more concrete form by asking how the hydrogen ion concentration is preserved at almost a constant. In this form, the nomogram provides at least a partial answer. For it reveals precisely what happens when an increase in one variable component, which one would expect to upset the hydrogen ion concentration, is introduced. Because of the twenty triadic relations it not only alters the hydrogen ion component but also affects the other four variables in such a manner that they act upon the hydrogen ion concentration to compensate any upset. Thus, although chemical decomposition and physical alterations are always taking place, the factors possess such physico-chemical properties and are so intricately interrelated that processes either run in cycles to repeat themselves, or any disorganizing chemical change is compensated and counteracted by the very factor which initiated it. In this fashion nature has reared temporarily stable wholes out of chemical change and has gained systems with dynamic attributes because of the release of stored energy which internal chemical decomposition permits.

The nearest analogy to its procedure is to be found in that of a cowboy on the Western plains when he is faced with a stampede. Every fall he goes up into the higher regions of the Rockies to bring down the cattle before coming storms cut them off. During the long drive home, the steers often become nettled by the continuous peppering of dust and pebbles which ceaseless gales hurl against them. Under such circumstances they become restless under restraint and often break wildly from control. There are two possible methods of dealing with such a situation. One is to calm down each individual animal so that it will live in a relation of Christian brotherliness with the other members of the herd. This is the method that nature may have used on the molar level in producing most gross inorganic objects, such as the stone. The whole may be held together because of the intrinsic chemical properties of the constituent atomic or molecular materials. But the cowboy has found the task of individual conversion too laborious an undertaking to be efficacious with his stampeding steers. Instead, he follows the other alternative. His pony is sent in a wide circle around the group turning one critter in here, another in there, until he has them playing upon and counterbalancing each other, with the result, at the end of his labors, that each steer may be running as fast as ever but the whole herd is standing still. Nature has used a method analogous to this in producing living organisms.

But nature is not a cowboy. Nor does it possess a cow-pony. How has it accomplished this result? This brings us to the really basic question. What must be the basic nature of a universe which can produce such a system by such methods? In other words, what first principles of science does the living system, revealed in Henderson's nomogram of the blood, imply?

We can say immediately that nature must be kinetic atomic in character. We have but to combine Parmenides' arguments given in the first chapter with the two obvious facts that a living thing is physical and involves change to know that this is the case. The essentially chemical character, not only of the system as a whole, but of its apparently teleological behavior and dynamic attributes, which the nomogram reveals, makes this conclusion inescapable.

Nevertheless, the type of physico-chemical system which is present in a living organism cannot be produced by the traditional kinetic atomic theory. The latter theory is necessary for its existence but not sufficient.

According to traditional physical and chemical theory the stability and permanence of a system can have but two sources. Either it must be due to the physical attraction or chemical affinity of the microscopic constituents, or it must represent a statistical type of dynamic equilibrium. The dynamic type of equilibrium in the living thing and the fact that the source of its stability is not internal eliminates the first possibility.

Another factor which the nomogram reveals to be essential in life eliminates the second. We have noted that a living thing maintains itself because the possible normal deviations of any of its major physico-chemical components introduces automatic compensations for any disorganizing effect which it might produce. An examination of the blood system reveals, however, that this remarkable self-regulatory mechanism has not been built out of the ordinary immediately available physico-chemical materials of the inorganic environment. A very complicated chemical compound, haemoglobin, has been introduced. Moreover, Henderson has pointed out [12] that this compound serves three entirely disconnected functions, and that the necessary compensatory reactions would not be present were this not the case. In other words nature has produced living things by dropping into the flux of combining and disorganizing chemical materials which inorganic nature left over on our molar level, a very complicated substance with precisely those many chemical properties which would cause the flux to order itself into a dynamic permanence with such an intimate relation of its several components that any normal environmental influences upon it automatically compensates destructive effects to preserve the whole. In other words, the type of dynamic equilibrium which we find in living things involves a very improbable type of permutation. This eliminates the possibility of accounting for the living organism in terms of the traditional kinetic atomic theory, on statistical grounds. Traditional statistical principles could account for the origin of such a system, but since it involves an extremely improbable type of permutation it would disappear as soon as it arose. This follows because no lasting dynamical equilibrium is possible on traditional physical principles unless it represents a probable type of permutation of elements.

We find ourselves forced, therefore, by the physico-chemical character of living things which the nomogram unequivocally reveals, to amend our traditional kinetic atomic theory. Life is a physico-chemical system calling for a kinetic atomic theory, but it involves a type of physico-chemical stability which cannot be conditioned solely by traditional atomic principles.

What must the nature of this amendment be? An analysis of the essential factors involved in biological stability should provide an answer to this question.

An examination of the dynamic heterogeneous physico-chemical equilibrium which is a living organism, reveals that its stability rests upon three essential factors: (1) an external pole over against the internal pole of the equilibrium; (2) the constant intimacy of relationship of the six major components during chemical change when there are no chemical bonds to hold them in interrelationship, and (3) the presence of improbable extremely complex organic compounds with precisely those chemical properties which are necessary to automatically compensate all normal changes so that the hydrogen ion concentration is preserved at practically a constant.

Consider the second factor first. Without this intimacy of relationship the twenty triadic relations between the six major components of our system would not exist and the delicate adjustments necessary for the preservation of such a system would be out of the question. The dynamic character of our system, together with the continuous chemical change within it, indicates that this intimacy of relationship between the ultimate physico-chemical components cannot be due to bonds of chemical affinity. The first inclination is to refer this intimacy to the pressure of the earth's atmosphere, but since this is explicable apparently, in terms of traditional principles and we know that our organic system is not, such a procedure merely moves the difficulty into another science where it will be ignored, instead of solving it. The only safe procedure seems to be to go back behind inorganic systems to first principles themselves. From the point of view of the ultimate kinetic atomic level of nature, what must be the character of the atomic entities if this fact of intimacy which is not due to chemical attraction or statistical distribution is to be accounted for?

There seems to be but one answer to this question. The congested or intimate relation between physico-chemical materials which we find in a living thing must be an ultimate fact. The basic atomic entities of nature must be forced to remain in a congested relationship to each other regardless of whether they are tied by bonds of electrical or chemical affinity or not. Such bonds merely preserve permutations; they have nothing whatever to do with the density or congestion or intimacy of relationship of the atomic entities that are remaining in, or passing out of a given permutation. Only in this fashion can we explain why a physico-chemical system which is constituted of chemical changes, does not disintegrate. This new amendment to our traditional kinetic atomic theory, we shall call the principle of the constancy of the average density of nature. The atomic entities of nature are in a constant congestion or intimacy of relationship regardless of whether they are in stable or unstable permutations. Precisely the same idea was involved in Einstein's theory of the finite universe.

The essential rôle which such complex organic compounds as haemoglobin play in living systems gives us additional information concerning the first principles of science. The improbable yet permanent character of such compounds indicates that all stability in nature cannot be defined statistically in terms of the class of permutations which is the most probable. Statistical principles can account for the origin of such a system as a living organism, but not for its preservation. It follows, therefore, that the permanence of living things must depend upon intrinsic chemical properties and not merely upon the statistical distribution arising from microscopic atomic motion.

Once this is recognized and the principle of the constancy of average density is accepted, the existence of such physico-chemical systems as we find in living organisms becomes intelligible. Even with active chemical materials, nature is able to build stable systems. The motion of the microscopic particles is producing an endless series of permutations. The most probable combinations arise first. Apparently these represent permutations of elements that have chemical affinity for each other. These constitute ordinary local molar objects such as a stone. Other active materials remain. They enter into combination and pass out again producing a general effect of flux. With continuous mixture, most probably in water where life is supposed to have arisen, improbable compounds will arise, with time. Any such compound will enter into the complex of materials remaining after the inorganic systems are formed in a manner that will lead to its own destruction or to the preservation of the complex of which it forms a part. If the former is the case it will disappear and another will take its place; if the latter, then it and the system which results will be automatically selected. Thus, providing the conservation of average atomic density holds, nature can build stable systems out of improbable combinations and unstable materials, with time. Furthermore, the more unstable the materials, the more complex and improbable will be the compound which is necessary to compensate disorganizing tendencies and produce constancy. It happens that the presence of carbon in inorganic nature and the capacity of carbon to produce complicated chain compounds makes the automatic natural selection of the peculiar compounds which are necessary to build permanence out of chemical change fairly easy. It appears, therefore, that biological stability and organization has its basis in two different characteristics of the microscopic atomic entities. One is the intimacy of relationship between them which prevents the universe from going to pieces even when bonds of attraction do not hold them together, thereby ensuring a continual series of new permutations or chance variations. The other is the intrinsic physical and chemical properties of the atomic entities which cause certain permutations to remain and others to be replaced by a successor.

But what do we mean by intimacy of relationship between the microscopic particles, or by the principle of the conservation of average density? In a physical theory of nature one cannot maintain moving particles in a congested state with a formal principle. The direction of their motion must be changed. An external pressure must be imposed upon them. For this only an external force having its source in an encircling physical object will suffice. The simplest physical object capable of meeting these requirements is a huge hollow spherical atom surrounding all the traditional moving microscopic atoms of the whole of nature. Thus we find biological evidence giving us an entirely independent argument for the existence of the macroscopic atomic theory.[19]

If this atom is present to constantly congest the microscopic atoms so that complex nature cannot disintegrate even if unstable permutations arise, then nature can build the kind of temporary stability found in living things out of chemical flux and the rare compounds to which new permutations and time give rise. Certainly if there is one permutation, no matter how rare, which will drop into chemical flux a complex compound with precisely these properties necessary to compensate all normal destructive influences upon the resultant physico-chemical system, given time enough it must arise, and if stability depends not on probability but on an equilibrium between the microscopic and macroscopic so perfectly adjusted that it does not lead to its own disintegration, then having arisen it automatically selects itself to remain.

We find ourselves furnished with a basic atomic meaning for the Darwinian ideas of chance variation and natural selection. The former idea has its basis in the kinetic atomic principle which tends to produce all possible permutations; the latter, in the macroscopic atom, its congesting influence, and the specific chemical properties of compound substances, which cause rare permutations and their complex compounds to be preserved providing they do not give rise to physico-chemical changes which lead to their own destruction.

Our conclusion concerning the foundations of biological systems can be reached in another way. Consider the temporary dynamic type of physico-chemical equilibrium between internal and external elements which a living thing exhibits. One of the poles of this equilibrium is the physico-chemical materials immediately within it. But where is the other pole? It is as if we see the key-stone of an arch and one column of its support, but not the other. We must pass out from the key-stone in its position, which the nomographic picture of the equilibrium reveals, to the other unknown pole. We are led to oxygen pressure, to the geological processes which determine the earth's atmosphere, to the astronomical processes which provide the organism's energy, until we come to the primary principles which move matter and energy about. There seems to be no stopping point until we get to the limits of the universe and arrive at the atomic entities which condition everything. And even there, when we examine traditional physical theory, we find nothing but the microscopic atomic factor which we had before we started. Only one conclusion seems possible. There must be another physical principle standing over against the moving microscopic entities to congest them and cause them to select out complex compounds to form a dynamic equilibrium. Only in this addition to our traditional atomic theory do we find the missing second column of our arch, the external pole of organic stability. Otherwise, as Haldane has suggested, the physical theory of biological organization leaves us, like a cat chasing its tail in an interminable circle, always just on the verge of reaching our goal, but never actually attaining it.

The living organism is a physico-chemical system localized in visible molar form but containing within the very tension of its constitution, all the microscopic atomic entities of the whole of nature in equilibrium with the congesting macroscopic atom. It is because the later pole of this equilibrium is an eternal entity present in the future as well as the past, to which the microscopic processes which enter into life must adjust themselves, that organic nature exhibits a teleological character, notwithstanding the fact that it is conditioned by purely physical principles. Thus the baffling problem of mechanism and teleology is resolved. Life appears to be governed by a final cause because the macroscopic atom to which all stable systems must adjust themselves is an eternal entity and hence is in the future as much as the past. Moreover it is because the living organism is an equilibrium between the constituent physico-chemical materials within and a congesting organizing physical principle operating from without that it exhibits an organization which is not completely explicable in terms of its constituent microscopic materials. A solution of the problem of biological organization is at hand which does not force us to reject the physico-chemical categories that modern science has established.

Furthermore, the amendment to traditional biological theory which is necessary to reconcile the mechanical and physico-chemical nature of living things with their organic character, is precisely the one which is required in relativity theory to reconcile the physical and kinetic character of nature with its general metrical uniformity. Thus biology and physics are brought under a common single set of first principles.

References and Bibliography

1. Hippocrates, The Genuine Works of. Trans. by F. Adams. Sydenham Society.
2. Empedocles, The Fragments of. Trans. by W. E. Leonard. Open Court.
3. Aristotle. Historia Animalium. Eng. trans. Ed. by W. D. Ross. Vol. IV. Oxford Press.
   De Partibus, de Motu, de Incessu, & de Generatione, Animalium. Ibid. Vol. V.
4. C. Darwin. The Origin of Species.
5. W. Harvey. The Motion of the Heart and Blood in Animals. Dent.
6. Lavoisier, Oeuvres de. 6 vols. Paris.
7. C. Bernard. Experimental Medicine. Macmillan.
8. H. Driesch. The Science and Philosophy of the Organism. A. & C. Black.
9. L. J. Henderson. Blood: A Study in General Physiology. Yale Press.
10. L. J. Henderson. The Fitness of the Environment. Macmillan.
11. L. J. Henderson. The Order of Nature. Harvard Press.
12. L. J. Henderson. Orthogenesis from the Standpoint of the Biochemist. Amer. Naturalist. 1922.
13. J. S. Haldane. The New Physiology. Pp. 35, 76. Chas. Griffin & Co.
14. J. S. Haldane. Organism and Environment. Yale Press.
15. J. S. Haldane. Respiration. Ch. XIV, p. 88. Yale Press.

16. J. S. Haldane. The Sciences and Philosophy. Doubleday, Doran.
17. J. Barcroft. The Respiratory Function of the Blood. Pp. 16 & 65ff. Cambridge Press.
18. F. S. C. Northrop. Rignano's Hypothesis of a Vital Energy, etc. Journ. of Phil. XXIV. No. 13.
19. F. S. C. Northrop. The Problem of Organization in Biology. Thesis. Harvard University Library.
20. W. M. Bayliss. The Principles of General Physiology. Longmans Green.
21. D. Burns. An Introduction to Biophysics. J. & A. Churchill.

CHAPTER V

MAN

MAN is a rational animal. The sciences which concern themselves with the nature of the rational are formal logic and pure mathematics. Hence, an adequate account of the nature of man must consider the findings of mathematics and logic, as well as biology and physiology. In addition to being an animal, and in part, at least, rational, man is also a scientist; that is, he has awareness and knowledge. The science which studies awareness as such is pure introspective psychology; that which investigates the basis and mechanics of knowledge is known as epistemology. It appears, therefore, that no one can pretend to possess even a most general and introductory conception of the nature of man until he has considered the physiological, logical, psychical, and epistemological phases of his nature.

Moreover, since man is but nature expressing itself in one of its parts, these five phases of man's nature cannot be separated from the basic entities and relations that condition nature's character and behavior. Thus the character of man as animal and scientist derives illumination from and throws light upon the first principles of science. To consider man apart from first principles is to sever him from the roots of his character, and to misunderstand both him and nature, and their common product, science.

Thus, we find ourselves facing the important question: What is the relation between the physiological, logical, psychical, and epistemological nature of man and the first principles of science? Since the psychical and epistemological phases of this question depend upon everything else, and are determined as much by the character of nature as the character of man, and amount to a summary of all that we have learned concerning both, we shall leave them for the next and concluding chapter. The present chapter will concern itself with the physiological and logical phases of our problem.

THE PHYSIOLOGICAL NATURE OF MAN

In the previous chapter we considered the specific character of the most general protoplasmic system. Henderson's nomogram revealed those general physico-chemical characteristics which man has in common with all other forms of life. Our present task is to determine in what the differences consist, which distinguish man from other living systems. This undertaking falls naturally into three parts: the raw materials of life,[1] the organization of those materials, and the mechanics of the evolutionary process which has made them what they are in man.

Heredity and the Theory of the Gene

The science of the raw materials of life is genetics. Its theme is heredity.

The significance of heredity is self-evident. It appears the moment one considers the relation between the characteristics of parents and their offspring. An unexpected principle reveals itself when this relationship is studied carefully and experimentally. It appears that the observed characteristics of living things have an atomic and permutative character as they present themselves in connected generations. A single example will indicate this. Take such a simple case as the crossing of red and white flowers. The product is pink. But in succeeding generations the red and white can be regained. The situation is quite analogous to the one which drove chemists to to the discovery of the atomic theory of matter. Lavoisier heated the semi-liquid silvery mercury with oxygen and gained a powder of an entirely different color, which upon being heated again with carbon gave back the original semi-liquid silvery mercury without any loss in weight. Similarly organic traits have this atomic and permutative character. We can understand, therefore, why the biologists, and particularly the geneticists, have been led to the doctrine of hereditary elements, called genes.[2]

Moreover, these atomic characters seem to relate themselves according to the permutative rules of chance combination. The laws of Mendelian heredity express this fact. Also the genes or hereditary elements located in the chromosomes of every cell are physico-chemical substances. The experiments of Muller,[3] in which they, and through them, the subsequent progeny, are affected by x-rays, establishes this point. Hence they have a complex physico-chemical nature. In other words, although they are atomic from a genetic point of view when one ignores their changes, they are complex from the physico-chemical point of view. Because of this complex constitution there is nothing to prevent them from undergoing an addition to, or reorganization in, their structure, such as Muller has produced. Such a modification in structure the geneticist calls a mutation. Since Muller's experiments, in which such a mutation is produced by x-rays, induce a modification in observable specific parts of the offspring, a definite connection seems to be established between the inherited characteristics of living organisms and the organization of certain complex physico-chemical substances located in the chromosomes of the germ cells. It seems to be by a splitting and combining of these complex substances and the chromosomes in which they are located that the physiological assets and liabilities which we take over from our two parents at the beginning of life are in part at least determined.[4]

Moreover, contemporary biologists are convinced that reorganizations of these complex substances are at the basis of the mechanics of the evolution of new forms of life. There can be little doubt but that more of a load has been placed upon the genes than they are able to carry. We shall find that such an explanation, while partially correct, is entirely too simple and easy an account, and ignores too many other known factors to be regarded as the whole truth. On the other hand, those who emphasize the positive causal influence of external factors in evolution must never forget one fact: It is not enough for the environment to produce a new organization in living matter; that new organization must be inherited before a new form can establish itself, and inheritance is tied up with combinations, mutations, and reorganizations in these complex substances in the chromosomes. Also, regardless of the degree of seriousness with which one takes the theory of the gene, one fact independent of all entities, whether postulated or real, remains; namely, certain hereditary traits of human beings, in common with all other forms of life, reveal in their relations that an internal, pluralistic, permutative atomic principle is one foundation of human nature. The manner in which human traits shuffle themselves from generation to generation, entering into and passing out of all possible permutations permits no escape from this conclusion. The pluralistic microscopic principle is at the basis of human as well as inanimate nature.

Nor does this surprise us. It is exactly what our first principles of science require. The macroscopic atomic theory necessitates that a kinetic microscopic principle must exhibit itself in the raw constituent materials of any complex natural system. Moreover, our analysis of the foundations of organic stability points toward the same conclusion. Recall again the delicately but dependably balanced physico-chemical system which our analysis of Henderson's nomographic picture of a typical protoplasmic system revealed. A living thing is a temporary dynamic physico-chemical equilibrium between external and internal elements, which is constituted of such peculiar complex chemical compounds, by the chance variations arising from the kinetic microscopic principle and the automatic natural selection due in part to the macroscopic atom, that the active chemical processes involved in life either run in a cycle to repeat themselves and preserve certain constancies through the system as a whole, or automatically initiate compensatory reactions for any destructive influence which the normal circumstances of life introduce. Obviously, the kinetic microscopic principle exhibits itself in the dynamic chemical properties of this system. But it is also present in a more specific sense. For the stability of the system and the type of reaction that it gives to a certain change in one of its components is determined not merely by the organization which the macroscopic atom imposes but also by the presence of certain complex and unusual physico-chemical compounds. With these key substances present, the normal materials of the inorganic environment, congested eternally by the macroscopic atom, must interact to produce those enlarging systems which we call growing living organisms; without them a flux of chemical composition and decomposition would cover the face of the earth. Moreover, since these compounds are essential for the existence of life it follows that they must be transmitted to succeeding generations, if a given form is to preserve its type. Furthermore, since organisms result automatically from the normal interaction of these key compounds with the environment, it is only necessary to transmit them to succeeding generations to preserve a given form of life. We can understand, therefore, why the science of genetics has found heredity to be centered in certain complex physico-chemical substances located in the germ cells.

Also, since these compounds have their basis in moving physical particles which tend to give rise to new permutations, we can appreciate why spontaneous, internally conditioned mutations seem to be one key to the mechanics of evolution, and why the observable hereditary materials of man's nature exhibit an atomic and permutative character in their relations.

Much has been written by defenders of a monistic philosophy concerning the failure of the physical theory of nature to do justice to the nature of man. Nevertheless, the plain fact is that the character and behaviour of the observable raw materials with which we begin life are unintelligible unless nature is a physical kinetic atomic system. More may be involved, but this at least is certain. The raw materials of human nature find their basis in the microscopic atomic principle of our theory.

Organization

But in addition to the raw materials of human nature there is their organization or form. Certain experiments in embryology indicate that the source of this is not to be found in the hereditary materials alone. At an early stage in the development of the fertilized egg of certain organisms, Spemann[6] and O. Mangold[7] have transplanted groups of cells which normally produce certain tissues and organs of the animal. They find in certain instances that the fate of the cells is determined, not by their internal constitution, but by their relation to other cells of the organism. For example, the portion which normally develops into the brain of a certain organism may be transplanted to another portion of the animal, where it develops into the quite different organs of that region. Evidently form, as well as the hereditary material, determines the character of a living thing.

What is the source of this form? Another experimental discovery made by Spemann throws some light upon this question.[5] He found a semi-circular area known as the upper lip of the blastopore of the developing embryo which behaves quite differently when transplanted. Instead of changing its gross character and fate, it retains its normal character and becomes a centre around which neighboring material arranges itself. When only part of this material is transplanted the remainder of the system arranges itself relatively to both parts. The orientation relatively to the normal position of this material, Spemann terms the primary embryo; the orientation relatively to the transplanted material, he calls the secondary embryo. Two points of significance for subsequent considerations are to be noted. In the first place, the secondary embryo need not have the same orientation as the primary embryo. And secondly, this region of the embryo which does not change its fate when transplanted is near, but is not identical with, the material that gives rise to the nervous system. In other words, although the nervous system does not arise out of this dominating material, it arises in the immediate neighborhood of this material.

This portion of the developing embryo which does not change its fate when transplanted, Spemann calls the organizer.[5] Is not this title somewhat misleading? The fact which it denotes is merely this: Certain portions of the embryo change their fate when transplanted, others do not. But to use the term organizer to denote the latter half of this fact, is to suggest that the material which does not change its fate, is the active source of organization. This is quite a different thing, and something concerning which, so far as one can see, his experimental evidence says absolutely nothing. One does not regard a pile in the moving water of a lake as the active cause of the wave formation which exists around it, even though the wave formation comes and goes with a shift in the position of the pile from one side of the lake to the other. The organizer may have no more significant rôle, or as is even more likely, its significant activity may center in active chemical materials, or integrative factors much more fundamental and far-reaching than it.

Moreover, in thinking about the developing embryo, it is well to remember that the diverse organs which finally arise in the more complex living systems are but crystallizations of certain original properties of protoplasm. Thus the circulatory system, with its heart and associated network of arteries and veins and capillaries, is but an expression of the flowing fluid character of original semiliquid protoplasm. Similarly, the metabolic system has arisen out of the chemical and energy transformations within this fluid material. Now, the specific organic structure which integrates the other processes and organs of the body to produce the unitary form and organization of the system as a whole is the nervous system. Hence a consideration of its essential characteristics and origin should throw light on the problem of form.

It has three major characteristics. In the first place, as we have indicated, it integrates [8] the organism as a whole. Thus we may regard it as a crystallization of the form of original protoplasm. Secondly, it is the center of sensitivity. This attribute of the nervous system is but an expression of that most general property of original protoplasm, known as irritability. From this the third important attribute of the nervous system arises. It records within the internal organization of the organism, the relations of the living creature to the rest of nature. In short, it is an equilibrium between the many raw hereditary materials within and the external factors without. This property is present in living matter long before specific neural organs arise. For example, the single-celled amoeba responds to objects within its immediate environment with which it is not in direct visible physical contact. Evidently, the specific character of the environment records itself, at least in part, in the internal organization of each living thing.

One would judge, considering the emphasis which they put upon genetics, that most contemporary biologists would place the source of the nervous system completely in the genes. But this does not seem to be justified. One of the essential functions of the nervous system is its rôle as an integrative factor. Integration calls for a unitary principle, whereas the genes represent a pluralistic principle. Moreover, the earlier experiments of Spemann suggest that form does not depend upon the internal bodily materials alone. It is something into which they enter, but not something which they alone produce. Also, the manner in which the nervous system records external relations and objects points quite unequivocally toward an external organic factor, rather than toward internal entities as its source.

Furthermore, the organic character of the inorganic universe and of living things has been so firmly established by recent physics and physiological chemistry, that one must be very suspicious of all theories of biological organization which would locate its source wholly in internal entities, whether they be entelechies, or genes, or organizers. At a time when certain physicists are threatening to throw all entities out of inorganic nature, and leave nothing but mathematical equations and formal relatedness, it comes with something of a shock, to find so many contemporary experimental zoologists, locating the form of that most organic of systems, the living organism, in nothing but internal entities.

One finds it difficult to escape the impression that contemporary biology, except for the work of the physiological chemists, is in much the same state as were physics and chemistry a decade ago, when everything was located in electrons and protons, and the wave theory and field conditions which supplement these microscopic particles had not been discovered.

Yet there is no reason why biological philosophy should be in such a state. Enough evidence is at hand to correct the one-sided over-emphasis of the gene theory. The only difficulty is that one must cultivate the philosophical attitude of mind in order to know it. But this is inevitable. The chemist is concerned with the properties of gross matter, the physicist with matter and motion, the astronomer with stellar bodies, the student of thermodynamics with energy and its transformations; but the living organism is all these factors merged into a grand synthesis. Moreover, the essence of life centers in the cosmic forces which produce the synthesis, as well as in the local constituent materials themselves. Now, the science of the synthesis of the sciences is philosophy. Hence to understand life without looking at one's local technical experimental evidence from the philosophical point of view is impossible.

This does not mean that every experimental or descriptive biologist should close his laboratory and study professional philosophy; it does mean, however, that the present approach to living things through intensive analysis must be supplemented with the construction of a general accurate picture of the living organism in its actual physical, geological, chemical, astronomical and thermo-dynamical connection with the rest of nature. It is because Henderson's nomogram forms the starting point for such a construction that it is important. There is even more need for the theoretical biologist than for the theoretical physicist.

Otherwise the senses are always misleading the experimental worker. Unless one is continually correcting and supplementing observation with a correct conception of the general physico-chemical nature of life as a whole, which reveals it to be a complex heterogeneous physico-chemical equilibrium rooted as unequivocally in the physical and dynamical foundations of the environment, as in internal private genetical materials, the apparently solid character of living things causes one to regard them as a purely local entity. Then the notion arises that life is to be understood solely in terms of what is contained in the gross local body itself. Once this false assumption takes hold of the mind, the postulation of an endless series of entities both physical and vital is inevitable, and one's biological theory becomes increasingly false to the extent that one's experimental evidence is increasingly sound.

Because certain local observable inherited traits of man reveal an internal pluralistic atomic principle to be at the basis of his nature, one must not conclude that all his characteristics have such a basis. Nor must one conclude because certain transplanted materials of the developing embryo are unable to change their characteristics when transplanted, that they are the active cause of organization. Before an adequate theory can be expected we must consider the general characteristics of the living organism in the light of the first principle of science which condition them.

This brings us back to the general physico-chemical properties of life which were revealed in L. J. Henderson's nomogram of the blood, and to the theory of the first principles of science which our analysis of inorganic nature has suggested. We must bring our conception of the nature of man into harmony with these factors.

An analysis of certain strategic conceptions in relativity and atomic physics and general physiology has forced us to the discovery of a new theory of the first principles of science. Nature is constituted not merely of the electrons of traditional physical theory, but also of one large spherical physical macroscopic atom which surrounds and congests them. The congestion arises solely from the fact that the diameter of the macroscopic atom is finite in length and so short relatively to the very large finite numbers of microscopic atoms which it contains, that they are crowded together to form an equilibrium. This equilibrium between the outward pressure of the microscopic particles and the congestion of the macroscopic atom is compound, or molar, nature. It is to be noted that the macroscopic atom produces its effect by no mysterious or changing activity, but by keeping its properties fixed and merely being what it is. To use Aristotle's expression, it is an "unmoved mover". The fixed physical spherical form of this macroscopic atom necessitates that the equilibrium which is molar nature is given a form. Tradition has termed this macroscopic form, the order of nature.[9] Moreover this equilibrium between the microscopic and the macroscopic must have a dual character. The many enclosed moving microscopic atoms constitute an arithmetical, pluralistic, internal, and kinetic principle, and hence tend to produce many different changing systems; the one static, enclosing, spherical, macroscopic atom represents an externally imposed geometrical, unifying, and eternal principle, and hence tends to produce one static ordered system. Thus it does not surprise us that stellar nature is more static than biological nature, or that nature as a whole is one static and more eternal order of many local temporary organisms.

This must be the case if a changeless, perfectly formed, encircling, geometrical one, and a contained, moving, arithmetical many are in polar opposition.

Everywhere we have found this opposition and synthesis of polar principles to apply. Space-time is a union of general macroscopic metrical uniformity and constancy, and local microscopic heterogeneity and variability; atomic physics and optics exhibit a union of a continuous monistic wave medium with its macroscopic boundary conditions, and a discontinuous and contingent grouping of particles; and the most general protoplasmic system, when viewed from the physico-chemical point of view, appears as an organized equilibrium of the atomic materials of the whole of nature, including those without as well as those within the epidermis of its body. In short, everywhere in inorganic nature we find externally conditioned organization and continuous field conditions merged with local, internally conditioned, complex materials and the discontinuity which they entail. Out of nature has come the living organism and from its first comparatively simple and homogeneous form has come man. Hence, it would be strange indeed if this opposition and synthesis of polar principles did not apply to him also.

The presence of the microscopic principle has already been discovered in the dynamic chemical attributes of life, in its dependence upon very complex key compounds for its stability and evolution, and in the atomic and permutative character of inherited characteristics.

But our theory of first principles has other consequences. In addition to the kinetic microscopic principle, there is the encircling macroscopic atom. Its spherical geometrical character adds a formal factor at the basis of natural phenomena, and what is more important still, the embodiment of this form in a physical atom insures that its presence is effective; it has the physical forces at its disposal, which are necessary to change the direction of motion of the moving microscopic particles. This form must exhibit itself immediately in the general order of nature, and because of the congesting factor, this order of nature must exhibit itself in every local system. Hence, upon the raw materials of life there must be imposed organization. We are not surprised, therefore, that Spemann found the fate of certain cells in the developing embryo to depend upon their relation to other cells of this system, or that such competent students of life as Aristotle, Claude Bernard, and J. S. Haldane discovered living organisms to be physical systems whose organization is not intelligible in terms of their internal constituent materials alone. There is the one macroscopic order imposed from without as well as the pluralistic microscopic permutative principle operating from within.

But if the living organism is the sensitive physico-chemical system which Henderson's nomogram has revealed it to be, and if its organization is conditioned by the macroscopic atom operating through the order of nature from without, it follows that there must be an equilibrium and order between any organism and other organisms, and the rest of nature, as well as within it. In fact, the local organization of a plant or animal can have no meaning apart from the one wider order of nature as a whole. L. J. Henderson has found this to be the case,[10] for he has discovered a most intimate and essential relation to exist between the properties of living organisms and the most extensive and deep-rooted properties of the environment. If this be so, the general macroscopic relations which join an organism to the rest of nature, and the specific local relative relations which join it to the objects of its immediate environment must be to some extent internal to it, and must exhibit themselves in the form and specific internal physiology and anatomy of the organism itself. Such externally conditioned internal organization exists in living things. In the amoeba it appears in a capacity to sense the presence of objects with which it is not in direct physical contact, in man it exhibits itself as a definite set of organs known as the nervous system.

It appears that one has but to carry the polar principle of the macroscopic atomic theory to its consequences in such a delicately and sensitively balanced physico-chemical system as Henderson's work has revealed living matter to be, to discover that peculiar merging of raw hereditary materials with an externally conditioned form, which is the physiological nature of man. It appears also that the physico-chemical equilibrium which is man, is dual in character. There is the balance between internal and external factors revealed in Henderson's nomogram, and there is the equilibrium, between this balance and other organic and inorganic systems, which is the nervous system and its externally and internally grounded internal organization and control. It is because of its localization in the nervous system, which is grounded as much in nature as in man, that man's consciousness is centered as much, and usually more, on nature than on himself.

The importance of the nervous system is such that we must examine its origin and development more in detail. This brings us to the very important work of the American physiologist C. M. Child.[11] His experiments show that external factors, because of the greater effect which they have on certain portions of the organism than others, lay down regional differences in the system, and tend to orient the formal arrangement of the internal material relatively to these differences. Let a ray of light fall upon a simple protoplasmic system. It affects one portion of the system more than others. Naturally this makes a region of higher activity in that part. This sensitivity gradually works out through the system growing weaker and weaker as it spreads and proceeds. Thus different levels of activity are created. This series of regional differences Child calls a physiological gradient.

He has been inclined to identify these differences with different rates of metabolic activity. Several considerations suggest that this may need amendment. It would seem as if metabolic activity should depend more on internal chemical materials and the localization of especially active compounds. Also acording to Child, the gradient is the incipient nervous system. Now the nervous system is like a telephone system rather than a high tension power transmission system. The energy of the body is not stored in the brain and then sent out along the nerves to the organs where it is needed. Instead, it is located throughout the body and merely set off by very small amounts of energy sent through the nerves. Hence, one would not expect all the energy of the body to be lined in a series of decreasing quantity as one proceeds down the gradient. Furthermore, our theory of first principles suggests that the effect of the external factor is to introduce congestion so that organization is inevitable and to impose formal differences. To be sure, form can only be present if it is embodied in physical factors, as Child has correctly noted. But the physical differences necessary to embody a formal factor can be conditioned by a differential distribution of but a portion of the energy of the body along the gradient, without assigning all the energy of the body to such a distribution. However, this is a minor point. In any event, the essential point in Child's theory remains true. External factors condition regional differences of a physical character in living things and these at least contribute to the determination of the symmetry and form of the body.

Child's work also suggests that these gradients are independent of the specific nature of the external stimulus, and that they arise only in relation to external factors. There seems to be no reason why this should not be the case, even if they orient themselves relatively to the materials which Spemann calls organizers.[12] When one places the fact of the external origin of the gradient, in conjunction with the additional evidence which suggests that it determines the axial form of the body, it appears that the available experimental findings support our theory of the form of living things at the crucial points in our doctrine.

This theory is that the specific form of a living creature is to be defined completely neither in terms of internal entities such as genes, or entelechies, or organizers, nor in terms of external environmental influences alone, but that it is the product of the merging of an internal kinetic pluralistic principle and an externally imposed single formal principle. The evidence for the pluralistic principle is present in the theory of genes and in the varying potentialities of the parts of the embryo when transplanted, as discovered by Spemann. The externally imposed unifying formal factor exhibits itself in Child's theory of the physiological gradient.

Of course Child's particular formulation of the manner in which the essential observable internal structure of an organism is determined in part by its environment pole is not essential to our theory. All that our theory prescribes is that the living thing is a physico-chemical system which includes the rest of nature, in part, in its very constitution, and is so intimately and essentially related to other similar organic and inorganic systems by the congestion introduced by the macroscopic atom, that the character of the environment must register itself in the specific internal and more local constitution of the local organism. We have chosen Child's theory because it is the only one, in experimental zoölogy, with any respectable amount of evidence behind it, that gives concrete and specific expression to this necessary fact. That his theory expresses the precise manner in which the character of the environment is registered in the observable local organization of the organism is a contention which may be open for debate, but that the kind of determination of organic characteristics for which it stands, exists, is something which no one can doubt, so far as one can see, unless he rejects all that physiological chemistry teaches us concerning the sensitive character and far-reaching environmental foundations of those physico-chemical systems which we call living organisms.

If this be admitted, we are prepared to summarize the steps which connect the specific organization of living things with the macroscopic atom, which is the ultimate unifying organizing principle in nature. Child's work indicates that the specific symmetry and form of living things is determined in part at least by the physiological gradient and that this gradient never arises except in relation to external factors. This places the one source of organization unequivocally in the order of nature, and calls for some source of that order. Our analysis of the spatial structure of this order forced us to locate it in the macroscopic atom. Thus there appears to be a very definite and concrete sense in which the macroscopic atom operates through the order of nature to condition the specific organization of living things.

Two other properties of the gradient must be noted. In the first place, the top of the gradient dominates the lower regions, and hence, if it is the only gradient, the other systems and processes of the organism. Secondly, Child has pointed out that in an axial type of organism such as man the controlling region of the nervous system develops at the dominant end of the major physiological gradient. In general, the nervous system first appears as a definite localized structure at or near the anterior end of the major gradient when the primary germ layers of the embryo have become differentiated and before other definite organs and tissues appear.[11] Spemann's organizer may be an expression of this, as Child has suggested.[12] Furthermore, experiments reveal that the structure of the nervous system and the direction of its growth is associated with the physiological gradient, and that the dominant portion of the central nervous system is located at or near the upper dominant end of the gradient. In man this dominant portion of the nervous system is the cortex of the brain. Thus in terms of the action of external factors upon the raw hereditary materials we can understand why man is an organized physico-chemical system with a nervous system dominated and coördinated by the cortex of his brain.

Certainly if such a system were conscious, its consciousness would be located extensively in nature as well as relatively to this particular body, since the nervous system is an equilibrium between the raw protoplasmic materials within, and external nature. Furthermore, the body itself is really an equilibrium involving the whole of nature. Thus, knowledge must have both an absolute and a relative character, absolute in so far as man and his nervous system has its basis in the whole of nature which is common to all local systems and standpoints; relative, to the extent that the purely local and atomic permutations record themselves in his neural constitution. Only by understanding the relation between neural organization and the general physico-chemical organization of man and nature [13] can one hope to understand knowledge.

A full understanding of the physiological nature of man is not possible without a consideration of the evolution and origin of new and more complicated forms of life. We have suggested how the nervous system may have originated to determine a given form of life. How do new forms arise?

The Mechanics of Evolution

The prevalent answer to this question is in terms of a philosophy called emergent evolution. It appears to provide a most plausible solution of the difficulty. Unfortunately, it possesses two fatal weaknesses. In the first place, its plausibility is only apparent. We are simply told that new forms of life arise because they emerge. Man asks for a specific solution to a complex problem and he is given a verbal sedative.[14] But emergent evolution has a more fatal weakness. It cannot be true. For it rests on the assumption that time is a primary concept. Relativity physics informs us that this is not the case. Time is relative and must be defined in terms of matter and motion. We must look elsewhere for a philosophy of evolution.

A more specific answer is given by the geneticists in

terms of the mutation of genes. New forms arise because the genes mutate. Undoubtedly this is part of the truth. For the possibility of life depends on the existence of certain complex physico-chemical compounds. It is reasonable to suppose, therefore, that the origin of new forms must depend upon their reorganization into new complex compounds. Furthermore, new forms cannot establish themselves unless these complex physico-chemical materials in the germ cells pass them on to the next generation.

But to know that a reorganization in the chemical components of the germ cells is necessary for the evolution of new forms is not to have proof that it is sufficient. In fact, we have very definite evidence that it is not. For if the hereditary materials are not sufficient to account for the organization of an existent form, it seems to follow quite necessarily, that they cannot be sufficient to account for the origin of new forms. Moreover, the attribution of everything that occurs in evolution to the mutation of the genes, while partially correct, as far as it goes, is entirely too easy a solution of the difficulty. It is too much like saying that it is due to abracadabra.

Specific knowledge is at hand which enables one to get a very much more accurate picture of the mechanics of evolution than this. Our previous remarks on the secondary status of the concept of time suggests the correct theory. Evolution is a temporal process. Since physics has revealed that time is relative and must be defined in terms of matter and motion, it follows that evolution must find its basis along with other processes in a physical and kinetic atomic theory. In fact, anyone who has sensed the physico-chemical character of life knows that this must be the case. Electrons and atoms cannot be the fundamental biological and physical concepts that they are if process is primary. Even Aristotle who accepted a philosophy of becoming saw this. Furthermore, if the problem in evolution is the origin of new forms, and form has its basis in an interaction between the raw hereditary materials within, and the environment without, it follows that we must look to the environment as well as the gene for the key to the process of evolution. Moreover, since the environment organizes the hereditary materials through the physiological gradient and the nervous system, it follows that the problem of evolution narrows down to a question of gradient and neural reorganizations with changing environmental influences and spontaneous or externally conditioned mutations of the genes. Certainly the science of geology assures us that such environmental changes occur. Moreover, Pike has been driven, by his own experimental findings in connection with the nervous system, to the conclusion that it is a most essential key to the origin of new forms in evolution.[15] In fact, the writer is greatly indebted to him for the first awareness of this aspect of the problem.

But traditional biology could accept this solution only by forgetting its physics. For Pike has also reminded us that the genes cannot be reorganized, nor can new neural organization be brought into existence without work being done upon them. This calls for energy. If the second law of thermo-dynamics is a literal temporal principle, what is the source of this energy? One can point to the fact that it comes from the sun. But this merely shifts the difficulty out of biology into astronomy where it is usually ignored. Again we see how futile it is to talk about life without the philosophical attitude of mind which brings the findings of one special science into conjunction with the ideas of another. We discover also how important our analysis of the second law of thermo-dynamics, and our new theory of first principles, is for biology. For both reveal that the second law is a physical and logical principle rather than a temporal process. And the macroscopic atom insures that there is a basic principle in nature inducing organization and providing spontaneously for the organization of new forms, as well as the destructive disorganizing principle which Carnot discovered.

Given this source of energy, and changes in the environment or the genes, the reorganization of the nervous system follows. And considering the most organic character of living things, which Henderson's nomogram has revealed, it is unreasonable to suppose that radical environmental changes producing these radical internal reorganizations would not affect the complex organic compounds in the germ cells, especially since they are the key to the stability of the system. Certainly the organic character of living systems and nature puts the burden of proof on those who would assert the contrary. However, it also follows that no amount of external change will produce a new form unless complex compounds arise, in the germ cells, which have precisely those properties necessary to enable the hereditary materials to grow by interaction with the environment into a new type which is stable. This is the truth expressed in the theory of the gene. As Henderson's nomogram indicated, not merely coördination but also complex chemical compounds are the key to the existence of life. Hence in evolution reorganization in the germ material and environmental changes must go together. Evolution as well as organization has its basis in a merging of an external form-producing principle and an internal pluralistic kinetic microscopic principle.

Let us state this doctrine in more concrete terms. Consider that stage in the process of evolution when the animals left the water for eventual habitation on dry land. The problem which nature faced, if forms were to survive under the new environmental conditions was to devise a new method of supplying the animal with oxygen. This problem was finally solved by laying down the wet membrane in the lungs between the circulatory system and the external atmosphere. Pike has indicated that such changes entailed the introduction of new nerves and a radical reorganization in the nervous system,[16] in which the center of control is shifted to higher and phylo-genetically newer regions of the central nervous system.

How did this occur? To say that the genes mutated is

certainly too easy an answer. Energy is necessary and it depends on external cosmic, rather than internal, local, factors. Furthermore, the nervous system is partially conditioned by external factors. There must be very powerful cosmic forces behind so mighty a reorganization. If the macroscopic atom is present to insure organization and is congesting the microscopic particles, certainly an externally imposed force working always for organization is present. Furthermore, with a changed environment acting upon such a delicately balanced and sensitive physico-chemical system as a living organism, it seems reasonable to suppose that chance, or even possibly, conditioned mutations might combine with external influences after many failures to produce a new organization. Certainly if there is a possible reorganization which will bring temporary organic equilibrium out of chemical flux, it seems inevitable considering the energy and time available, and the guidance of the physical principle of least work, that it must arise. Also when one recalls that external factors lay down physiological gradients, and that gradients determine the direction which growing nerve tissue pursues, changes in the reorganization of the nervous systems are not so hard to understand.

There are two types of fact which make this account plausible. One appeared in connection with some experimental investigations of my friend and colleague, H. S. Burr.[17] Desirous of increasing the number of olfactory nerves for a certain experiment which he had in mind, he added some olfactory nerve tissue to the normal supply of a certain organism. He was surprised to find that in over fifty per cent of the cases the transplanted tissue did not grow along the normal olfactory nerve path. Instead, it wandered off through the brain case into the central nervous system by an entirely original route. He suggests that the explanation is to be found in Child's theory of the physiological gradient. Due to changed conditions at the time when the experiment was performed, the distribution of gradient differences was not the same as when the original olfactory nerve was laid down. Thus a new physiological gradient arose to determine a new path for growing nerve tissue to follow. In any event we have experimental evidence that nerve tissue can develop in mature differentiated organisms along entirely new lines.

A second relevant fact has been discovered by Coghill.[18] He finds that the growth of a nerve fibre through a distance no greater than one-hundredth of a millimeter is adequate to produce a change in neural organization sufficient to make the difference between a swimming and a walking animal. In other words, all that the environment needs to do in order to transform the neural structure of a swimming into that of a walking organism is to establish a difference in potential on a region no greater than one-hundredth of a millimeter in length. It is evident that very marked changes in neural organization can occur without any undue demands being made upon the principles of our physical theory. Both evolution and organization seem to be the product of an internal pluralistic permutative or mutative principle and a unitary externally imposed formal principle.

Of course, the traditional biological philosophy has suggested this idea in its doctrine of an environmental selection of the materials which mutations produce. The uniqueness of our theory centers in the new definition of environment which it provides. According to traditional theory, environment, when defined in accordance with traditional assumptions, meant nothing more than the mere sum of all the local systems of nature; it was an effect of microscopic atomic motion, and in no primary sense a real cause. According to the macroscopic atomic theory environment is the sum of all the local systems plus an order and organization imposed upon them by the macroscopic atom. Thus the environment is not a mere effect of atomic motion, there is an element of the causal in it also. Hence, it conditions new forms as well as selects them from chance mutations. This also gives a basis for the notion of selection in the doctrine of natural selection. Although traditional biological philosophy used the latter doctrine, it was never really entitled to it; accepting only the microscopic atomic principle, it had but the principle of chance variations. Failure to define what it meant by environment covered up this fallacy.

An additional characteristic of the nervous system must be noted before we have gained a complete general picture of the physiological nature of man. It appears that new forms arise in the process of evolution by shifting the center of control to higher and phylo-genetically newer regions in the central nervous system. Hence, man's cortical mechanism, coming, as it does, at the present end of a long series of reorganizations, represents a rather complicated structure.[19] In the reorganization which has developed nature has taken two rather opposite directions. On the one hand, it has gained a certain amount of rigidity. This has been absolutely essential to preserve life. It exhibits itself in the constancy of the hydrogen ion concentration, in the control of respiration, and bodily temperature, and similar factors. On the other hand, nature has produced great flexibility in man. Without this, learning would be impossible and our educational institutions would be useless. To gain this it has been necessary to break away from the simple reflex with its specific response for a specific stimulus.[20] Many types of response are ready in waiting for any emergency. The advantage of such a neural mechanism in the struggle for existence is obvious. An organism cannot lay down a new neural organization when a harmful stimulus appears. Hence the organism possessing a wide repertoire of modes of response has an advantage. Thus we find that the two polar principles—pluralistic permutability and monistic rigidity—exhibit themselves in the central nervous system of man. In him both the microscopic and the macroscopic foundations of nature come to articulate expression.

Such is the physiological nature of man. Born of a synthesis of atomic opposites, he is a mixture of constancy and change. Grounded in the private compounds of his chromosomes and in the foundation of physical nature, he is at once here and everywhere. Mid-way between the melting pot of formless microscopic atomic flux and the mold of eternal rigid macroscopic physical form, he exists with one root of his nature in time and the grave of atomic dissolution and decay, and the other in eternity and the constancy of the form and order of macroscopic nature.

THE LOGICAL CHARACTER OF MAN

Greek science defined man as a rational animal. Modern science has been quite sure that man is an animal, but not so certain of his rationality.

Nor can we wonder at this. If nature and man are but accidental by-products of the motion of the microscopic atoms of traditional scientific theory, nothing else can be the case.

A clear conception of what we mean by the rational is necessary to make this evident. Most popular conceptions of the nature of rationality are unsound, and their unsoundness but echoes the insufficiency of modern thought. Facing the utter inadequacy of the traditional physical theory of first principles when it is confronted with the rational character of man, certain modern thinkers added a non-physical entity to the matter of scientific theory, and regarded this mental or spiritual substance as the source of reason. In some mysterious fashion this entity exhibited a property or faculty known as rationality.

It has always been a mystery why so rational an attribute of man should have so irrational a source. But defenders of this view had little time to reflect upon this difficulty, as they were continually engaged in trying to explain how a non-physical mind can be connected with and control the collection of moving atoms of a physical body. However, the error in this philosophy centers, not so much in the insoluble problems to which it gave rise, as in the false notion of the nature of the rational upon which it rests.

The Rational

There is nothing elusive or mysterious about the rational. No speculations concerning its nature are necessary. Long ago Greek science discovered that it is to be identified with form, and ever since then two sciences, pure mathematics and formal logic, have been investigating its specific character. Once this is recognized, it becomes evident that the neglect of the study of logic in the science of psychology which would reveal the nature of mind, and in our system of education which would make men rational, is one of the scandals of modern life. It is as ridiculous to expect to develop the minds of young men without teaching them logic, as it would be to expect to develop their bodies without giving them physical exercise. Similarly, a science of psychology which would expect to do justice to the rational character of man without the knowledge of the nature of form, which pure mathematics and logic reveals, is as absurd as one which would expect to account for bodily behavior without a knowledge of physics. No theory of mind can pretend to be scientific unless it is in accord with the established findings concerning the character of mathematical and logical form. It is because of this that the work of men like Russell,[22] Whitehead,[23] Hilbert,[24] Sheffer,[25] Lewis,[26][1] and Wittgenstein,[27] on the foundations of mathematics and logic is so important. They are discovering the nature of the rational, just as the physicists are revealing the nature of matter.

In the history of science there have been three major fundamental conceptions of the nature of form. Moreover, these three alternatives seem to exhaust the possibilities: It is an entity, an attribute of an entity, or a relation between entities. Greek inorganic science with its emphasis upon geometry, Platonism with its objective ideas, and logical realism with its thesis that universals are real and exist by themselves, represent the first conception; the functional theory of nature with its doctrine that form is an attribute of individual substances, and the logic of Aristotle which conceives of the formal properties of any categorical proposition in terms of the predication of an adjective of a subject represents the second; [28] and the modern logic of relations which seeks for the essence of form in the properties of relations, rather than in the entities or classes or adjectives related, rests upon the third.

It is to be noted that the logic of Aristotle and the logic of relations agree upon the doctrine of logical conceptualism: Universals or forms are real but they have no existence apart from specific individual things. This is expressed in the logic of relations by the proportional function with its variables.[22] Although the formal properties of any system are contained completely in the properties of the relations, the variable is present to continually remind one that forms do not exist apart from individual constants or particulars which they relate.

Let us consider these three conceptions of form in the light of different theories of the first principles of science. Greek inorganic science was forced by empirical evidence to the conclusion that nature is a system of ideal nonphysical geometrical spheres. Hence, its emphasis upon geometry and its conception of forms as entities is easy to understand. The Greek notion that rationality, which is the mark of divinity, is located in inorganic nature, and that man converts his instinctively irrational nature into a truly religious rational one, by learning to define his terms, and by studying mathematics and astronomy, also becomes intelligible.

Aristotle, however, was a biologist. The notion that reality is purely mathematical and formal is quite plausible for those who think about nature in terms of the distant expanses of astronomical space, but this becomes a very difficult theory to defend when one is dealing with the slimy body of an angle worm. The latter experience should be enough to convince anyone, as biological creatures convinced Aristotle, that there are material as well as formal causes. Once this is admitted the thesis that form is an attribute of a substance follows necessarily. Hence, it was inevitable that an age such as the Scholastic period, which rested on the functional philosophy of Aristotle, should have no difficulty in believing in the rational character of man. Man is rational because no concrete individual exists without a formal cause in his nature, and form is what we mean by the rational.

With modern science the notion of form as a causal factor was rejected. In other words, the first causes or foundations of our universe were declared to be completely irrational. Form was regarded as a relation between moving atoms, rather than an ultimate cause of the order of nature; and as no relations between the atoms are necessary the rational character of man is reduced to a mere accidental by-product of a basically irrational process which is headed toward a completely irrational end. Formal causes were to be defined completely in terms of physical causes. Is it any wonder that men in the modern world have rejected the science of logic, and have been content merely to allow their physical reactions to express themselves in pragmatically determined additional reactions, or that the formal side of nature and man has been ignored? Such is the logical consequence of the inadequate theory of first principles upon which modern science and the modern world has rested. It is not an accident that the dominant philosophy of America, which is the only purely modern civilization, has been pragmatism.[36,37] And it will not be an accident that pragmatism disappears when the full import of contemporary scientific discoveries are grasped. The conceptual pragmatism of C. I. Lewis,[38] which is inspired by formal logic and pure mathematics, is a step in this direction.

Not only did the first principles of modern science define all form or rationality in nature and man in terms of relations between moving physical atoms, but it also maintained a thorough-going pluralism. Only the kinetic microscopic atoms exist. We can understand therefore why modern logic is but another name for the logic of relations, and why the tendency in modern mathematics has been to reduce geometry to arithmetic. If form is a relation between moving atoms, and only the many numerically different microscopic particles exist, this must be the case.

The macroscopic atomic theory changes all this. For it combines Greek and Modern conceptions and places them in polar opposition. Since the traditional microscopic atomic theory is retained, the logic of relations holds. Form is a changing relation between entities. It is to be emphasized, however, that it is the entities and their motion which determine the specific character of their relations, and not the reverse. Thus the logic of relations reduces to the properties of certain entities, the most important of which, in the case of the microscopic atoms, is motion. In fact, motion is the key to most of the difficulties and problems of science and philosophy. And the important thing which we have learned from contemporary physics about motion is simply this: it *is*. To use the language of logic it is an indefinable or primitive idea. It is not here or now for it is not defined in terms of space-time; instead space and time are to be defined in terms of it.

Were the changing relational forms which arise from microscopic motion the only forms which exist, we could know nothing but what is perceived in the immediate instant, and the skepticism of Hume, and the impossibility of science would be inevitable. There could not be even such a degenerate type of logic as pragmatism, for workability calls for forms which are sufficiently necessary to permit "practice" to give a verdict, and forms which are certain only for an instant do not meet this elemental requirement.

But in opposition to the contingent changing forms which the kinetic microscopic principle of our theory introduces, there is also the eternal perfect spherical form which the macroscopic atom imposes. The presence of this atom with its spherical form throws an entirely new light upon the foundations of logic and reason in man and nature. In it we have a form which is a cause of the order of nature and the organization and intelligence of man, rather than one which is a mere effect of microscopic atomic motion. In short there is an elemental of the rational at the foundation of things. The intelligence of man is not a mere accidental temporary by-product of a basically irrational process. Man is not fighting an irrational universe, as Huxley supposed. There is a basic principle imposing form and reason upon things. Moreover, this eternal changeless form imposes macroscopic connections which are necessary, on nature.

Furthermore, the form of the macroscopic atom is not merely eternal and, hence, necessary; it is also the form of an individual physical substance. Thus the fundamental tenet of the Aristotelian logic is in part confirmed. All forms are not relations between relata; there is one form which is an attribute of an individual substance. In this fashion our theory merges the modern logic of relations with the traditional logic of Aristotle, by placing them in polar opposition. Nature is a succession of changing relational forms merged with or opposing one necessary eternal form, which is the attribute of a physical substance.

The relation between matter and form in the macroscopic atom is important. It involves a conception which is unique in Western thought. Aristotle regarded form as an attribute of a substance which is also physical. But he accepted the principle of becoming of the functional theory of nature. This forced him to regard matter as the source of contingency. Moreover, he put the opposition between the rational and the irrational in this distinction between form and matter. Thus, according to him, irrationality exists because stubborn contingent matter prevents perfect pure form from perfectly actualizing itself in concrete individual things. This viscious dichotomy between body and mind has echoed down through Western thought to nail the soul of man to a cross by dividing it against itself. The macroscopic atomic theory avoids this. Because it rests on the principle of being, the real cannot change its properties and hence there can be no contingency in atomic matter; contingency has its basis, as contemporary physics indicates, in the relations between particles of matter or electricity rather than in a shifting of their properties. And because the macroscopic atom is physical, yet has a perfect spherical form, there is no opposition between matter and form. In it, perfect form is perfectly actualized in an eternal concrete individual physical substance. Metaphysics no longer necessitates that one must be dumb in order to be a good athlete. Body does not oppose the actualization of form; instead it is the sole medium by which intelligence can be actualized.

Necessity and Possibility

The peculiar merging of stupidity with reason which man exhibits has its basis in the polar opposition between the changing relational forms to which the microscopic principle gives rise and the one necessary form which the macroscopic atom imposes on complex nature. This polar opposition of formal principles is important. It provides a meaning within nature itself for the distinction between the necessary and the possibles. The macroscopic atom is imposing one eternal changeless form on nature. The effect of this form appears over macroscopic regions and constitutes that portion of the actual formal universe which is necessary. But the microscopic principle, because of the kinetic character of the microscopic atoms, is producing a series of changing relational forms. These constitute the possibles. It is to be emphasized that they are as actual as the necessary form which the macroscopic atom introduces. These relational forms which constitute the possibles appear immediately in mind, and, as current physics has discovered, in the microscopic regions of nature. Thus, whereas the effect of the macroscopic atom upon complex nature, which results from the compounding of it and the moving microscopic particles, is to keep nature what it is, the effect of the microscopic principle is to break down existent organization and produce other worlds than this one. It is because nature is an opposition and synthesis of these two principles that our actual universe and experience is a mixture of necessity and possibility. The microscopic principle with its succession of possible relational forms appears in mind as imagination. Thus art as well as science, the pure mathematician who studies the formal structure of all possible worlds as well as the experimental physicist who studies this one, find their basis in nature. Pure mathematics is dealing with the actual objective universe as much as is experimental physics.

Imagination and possibility are not spurious afterthoughts created by that much over-worked entity the soul; instead they are an inevitable expression of the physical, formal, and atomic nature of things. Any natural system with unity or organization which is sufficiently fluid to give the kinetic atomic principle free play must have imagination. The most fantastic unrealistic work of art is a product of physical nature. Genius is rare because the neural organization which is sufficiently fluid to permit originality, yet sufficiently rigid to organize and control and refine the novelty that is discovered, involves a delicate balance of opposites which nature has difficulty in attaining. In most minds either the radical or the conservative principle predominates. Hence so many people are either extremely dull, or wildly fantastic, or a muddle of both with neither consistency nor convictions.

The inseparable connection between matter and form which our theory entails, has other important consequences. Less of a gulf arises between the order of observed nature and the formal structure of nature as conceived by scientific theory, than has been the case in traditional modern thought. Nothing is a more patent deliverance of observation than the sensuous form of things. To be sure there is chaos also, but, if anything, the chaos is more difficult to find than the order. Trees and even woods have symmetry of form. The ruthless struggle for existence is more difficult to observe. The stars in the heavens exhibit certain sensuous formal patterns, even though they are not the forms which the astronomer uses in his final classification. Everywhere observed nature exhibits form along with a submerged chaos.

The philosophy of Aristotle is the only system in Western thought which did justice to this fact. Platonism and the mathematical theory of nature gave an intelligible basis for the purely conceptual forms known by reason but were weak before the observed sensuous forms of experience. Modern science has been inadequate in its theory of both types of form.

As the modern man has moved over the hills and through the valleys of his planet he has observed form merged with chaos, and has often had his being lifted into the realm of ecstacy by the symmetry and delicacy of sensuous patterns that greet his eye. But if he has known the first principles of his science, he has had to deny this verdict of his senses. The chaos and tendency to disorder is genuine, his science tells him; the order, on the other hand, is counterfeit, for it is but a temporary accidental expression of a basically chaotic and non-formal process. At bottom all is but random atomic motion which, in the end, will convert all sensuous forms into a frigid chaos. In short, the only forms in nature are the changing relational forms produced by microscopic atomic motion.

Our theory puts an end to this obviously one-sided doctrine, by bringing scientific theory back into accord with the facts. The Greeks and leaders of Medieval thought were wrong in seeing little but order in nature, the Moderns have found a truth when they emphasized the presence of a chaotic principle at the basis of things. But chaos is not to be defined as an abstraction from or appearance of order, nor is order to be defined as a mere temporary by-product of chaos. Both are real, both are genuine. Nature is a mixture of order and chaos in which the order predominates over macroscopic regions, the chaos in microscopic ones, and both come to articulate expression on the molar level. This happens because the macroscopic atom opposes the many changing relational forms with one necessary unifying form. Moreover, because the form of the macroscopic atom is the form of a physical substance it exhibits its effects in observed physical nature. Thus the forms of the atomic entities known only by reason give rise to the sensuous forms of the world of observation. At last we have a philosophy which has the fertility of Aristotelianism with reference to the sensuous forms of experience while also admitting the basic contingent relatedness which modern science has discovered.

Our theory reveals its strength when confronted with the specific sensuous and conceptual forms of experience. Traditional modern science would define all forms in nature, whether conceptual or perceptual, in terms of the changing relational forms to which microscopic atomic motion gives rise. To be sure, microscopic motion has a formal character. Otherwise we could say nothing about it, and no consequences could be deduced from it. It has always been assumed that a rectilinear principle governs the motion of the microscopic atoms. In a consistent development of the physical theory this must be taken as an intrinsic property of the moving atoms. Otherwise we fall into the error, which relativity physics has corrected, of defining motion in terms of absolute space. This point is important for logic because it reveals that even the logic of relations which is to be identified with the changing relations to which microscopic motion gives rise, is to be defined in terms of properties of moving substances and hence reduces to a logic of an Aristotelian type. This indicates also that traditional modern science had a formal principle at the basis of its doctrine, notwithstanding its emphasis upon random physical motion as the cause of form. It attempted to define all form in terms of rectilinear motion and its compounding.

Two facts have always presented inseparable difficulties for this theory. The first is rotational motion. The second is the universal presence of the spiral.

A contemporary physicist has been reported to have said that if one analyzes the problems of physics to their foundations, one will always find rotational motion at the core of each difficulty. Even the students of the general theory of relativity have trouble in deciding what to prescribe concerning it. Jeans is convinced that a rotational force which is not defined in terms of simpler forces must be posited in astronomy.[38]

The universality of the spiral has impressed students who have taken art and nature at the face value of their materials.[39] From the Bohr atom, through molar nature with its swirling whirlpool, the growing goats' horn and climbing vine, to the astronomical nebulae, this sensuous form exhibits itself. One student of art [40] has gone so far as to say that all the proportions of Greek art are based upon it. Examine the details in any of Leonardo's drawings and one will rarely find a straight line. Certainly this great leader of the Renaissance recorded the nature of things. Such considerations are out of fashion in modern thought. In fact, the person who refers to them is usually looked at askance by contemporary scientists. But it is to be noted that this sense of superiority which our scientists exhibit toward this matter begs the question. For their attitude derives what validity it possesses from the modern assumption that random motion and a rectilinear principle is primary, and this is the very point at issue. Is it not time, in the light of the available evidence, to place our theory of the first principles of modern science into accord with the facts, rather than to define the observed forms away with an appeal to preconceived theory? When such a true scientist as Leonardo da Vinci reveals no all-sufficient primacy of rectilinear forms, and such a non-rectilinear form as the spiral exhibits itself without any regard to differences of material in which it is embodied, in every phase of nature from one extreme of its size to the other, is it not time to question the all-sufficiency of random rectilinear motion as a source of the observable forms of nature?

Our theory answers this question in the affirmative. Over against the random rectilinear forms of modern science it imposes a spherical form. The mean between these two seems to be the spiral. The spiral contains rotational motion within itself. Thus the two insuperable difficulties of modern science are met. All the order of nature is not a counterfeit. An irreducible macroscopic spherical form embodied in a physical substance insures the presence of eternal non-rectilinear forms in sensuous nature.

This opposition of formal principles exhibits itself in the science of mathematics as well. Ever since Pythagoras, this science has been facing the issue between continuity and discontinuity. This issue has exhibited itself in different theories of the relation between geometry and arithmetic. At different times in the history of science one of these sciences has been supposedly reduced to the other. In the modern world, the tendency has been to reduce geometry to arithmetic. But the reductions have never been decisive. Our theory indicates that such a desire for unity is an error. Both arithmetic and geometry represent primary categories. They represent forms in polar opposition. The microscopic pluralistic principle in the macroscopic atomic theory necessitates that numerical difference is an ultimate category, just as the spherical form of the indivisible macroscopic atom necessitates that there is an irreducible geometrical principle at the basis of things.

For the same reason continuity and discontinuity cannot be resolved, the one into the other. Quantum and wave mechanics has given us empirical evidence of this, but it also follows necessarily from our premises. The primacy of continuity means that the notion of the one cannot be defined in terms of the many. The Boundless of Anaximander is not the mere appearance of a basic discontinuous collection of unobservable microscopic atoms. For the macroscopic atom necessitates that complex nature is determined in part by the one, since its one encircling congesting form imposes a unity, and continuity, and order, upon the sum of the microscopic particles. But our theory indicates as unequivocally that discontinuity is not to be defined as a mere differentiation of a continuous one, or as a mere ideal limit reached from the one only by a method of extensive abstraction and suggestion. For there is the irreducible microscopic atomicity which the kinetic pole of our theory introduces. Nature is an opposition and synthesis of the one and the many, a conflict and merging of geometry and arithmetic.

Because of this, mathematics has its irrationals when it deals with complex nature and complex forms. Consider the diagonal of a right-angled triangle which has its other two sides equal. The numerical expression for the length of the diagonal $\sqrt{2}$. In determining the number of any length measuring is necessary. It involves the repetition of a standard unit of length. Thus the arithmetical and discontinuous exhibits its presence. But the actual segment of any complex form is determined by a non-discontinuous purely geometrical principle as well. Now the non-discontinuous and the discontinuous principles at the basis of nature and experience are externally related. In short, there is no logical or rational relation between them. Hence there is nothing to prevent two non-logically related principles from producing the irrational when they are merged in a complex system or experience. Are we permitted then to define the irrational as the merging of opposing externally related physical and formal principles?

Since the externally related formal principles which constitute the irrational are at the basis of physical nature, the irrational should exhibit itself in physics as well as mathematics. The peculiar merging of contradictories which current wave and quantum mechanics reveals, suggests that this is the case. Any complex object in nature both is and is not a wave, and is and is not a particle or collection of particles.

The Polar Principle and Human Nature

Out of inorganic nature has come man, and in the process of evolution this peculiar opposition and synthesis of necessary, externally-imposed order and internally-conditioned, changing relatedness has exhibited itself in him, as the first section of this chapter has indicated. Once this is recognized the peculiar nature of his rational character becomes intelligible. I refer to his natural capacity to think inconsistently as well as consistently. Certainly if there is one attribute of man's mind to which one would expect psychologists and philosophers to give most serious attention it is this. For what is more obvious than that natural man is a mixture of sense and nonsense. He thinks logically only after long training and discipline if at all. Yet one thinks back over modern thought in vain if one attempts to recall an adequate theory of this obvious fact. Even the one modern philosophy which prides itself on its theory of mind, the philosophy which proclaims that mind provides the key to everything else, rests on the coherence theory of truth. Could any theory be more out of accord with fact? For what does coherence mean but that reality is constituted of one internally related consistent system of forms.[34] This being the case and minds being real, how does it happen that man can have an inconsistent system of forms in his mind? Certainly, even the books of the absolute idealists indicate that he does.

This difficulty disappears immediately if one conceives of nature and man in terms of the externally related opposing formal principles which the macroscopic atomic theory necessitates. The macroscopic atom is imposing one necessary consistent system of forms on nature and over macroscopic regions, and, in part, in the mind of certain men at least it succeeds. Newtonian mechanics mistook this for the whole of theory, and came out with one necessary purely mechanical cast-iron universe. This macroscopic internal relatedness is the partial truth which has been mistaken by the absolute idealists for the whole truth. Both Newton and Hegel overlooked the existence of the possibles,—the changing relational forms introduced by the microscopic principle, which give rise to what Jeans has termed a loosely jointed universe. Since these temporary contingent forms can come and go and be merged with established forms or ideas in the mind of man, opposed and contradictory systems of forms exist in human minds.

In this difference between the actuality of the necessary and the actuality of the possibles, truth and error has its basis. But, viewed from the point of view of first principles, the relation between the forms of the possible and those of the actual, which defines falsity is as true and real as the relation which defines truth. Similarly, one must have a consistent theory of the existence of inconsistency. It is the great merit of absolute idealism that it recognized both of these facts. Its error arose because it used only one logic and defined the whole in terms of one internally related system of relations. Our theory would resolve these paradoxes concerning the truth of falsity and the consistency of inconsistency by distinguishing between two systems of logic. When we are talking about the consistency of a set of postulates or the truth of a given proposition we are asserting relations between relational forms. In fact, complex nature exhibits relational forms to us. Because they arise in two externally related atomic sources they may or may not be consistent systems. Similarly a form of the possible which we use to assert the existence of a similar form in the actual, may or may not have the same formal properties as the form of the actual. Hence truth and error arise. But these opposing relational forms are necessary consequences of forms which are attributes of atomic substances. When we get back to these primary forms we find no contradictions. The microscopic atom with its field is not both a wave and not a wave, and the macroscopic atom is not both a particle and not a particle. In short, there is no inconsistency whatever in conceiving of nature as constituted of many microscopic atoms with changeless compatible properties moving relatively to a macroscopic atom with changeless compatible properties. Nevertheless, given such a system, complex nature with a complex merging of relational forms in microscopic regions must exist. In this fashion our theory accounts for the existence of inconsistent forms in the mind of man without forcing mind and imagination and inconsistency out of nature, and at the same time provides us with a consistent theory of inconsistency. From this it follows necessarily that the factual test for consistency is not valid except for macroscopic regions. Inconsistency is impossible only in primary nature, or the world of being; in complex nature, or the world of existence, inconsistency exists. In fact, rare indeed is the individual who is not a walking contradiction.

The forms which constitute the ideas of men have two sources. Both the kinetic atomic principle which gives rise to all possible forms, and the macroscopic principle which tends to produce one constant complex of forms, operate in the human mind. Hence man's thought is both fixed and changing, exactly as his neural organization is a mixture of rigidity and flexibility. Without the kinetic permutative principle, learning and forgetting and memory would be impossible, and original thinking would be out of the question. But were this the only factor, man could know only what is in his consciousness now, the skepticism of Hume[32] would be the last word in philosophy, man's ideas would be continually changing, he would have no character, and would never be able to make up his mind concerning anything. More than one contemporary philosophy issues in such consequences. It is the static macroscopic principle which gives mind its determinateness, man his character, and places upon the temporal flow of forms the touch of the eternal which stops them long enough in their vague transition from one into the other, to reveal their character. As Socrates said long ago, pure flux or becoming would render conversation impossible.

In the light of our analysis of the foundations of the sensuous and conceptual forms of experience, it appears that forms are as objective and independent of knowing subjects as matter. Forms are not imposed on experience by the creative ego either necessarily as Kant[331] supposed, or pragmatically as Lewis suggests.[33] The formal or rational character of experience is as independent of man, as the physical. In fact, the mind of man, as the Greeks noted, is less rational than macroscopic inorganic nature. Man is partially rational because nature is partially rational, and man is one of nature's products.

Considering the polar opposition of physical and formal principles at the foundation of nature, is it any wonder that man, except when he is on his guard, or has undergone a most thorough scientific and logical training, is a most peculiar muddle of sense and nonsense? Only a brain caught in a very deep rut can prevent a most humorously incongruous set of ideas from arising at times. For nature herself is an inseparable identity of the formal and the physical, an interminable opposition of the necessary and the possibles, a peculiar conglomeration of the serious and the frivolous. This may not be the best of all possible systems of reality, but certainly it has left little out, not even the necessary.

Of such materials is man. Truly he is a physiological being in the original Greek sense of that term, a mixture of matter and form, of physics and logic.

With such a constitution he enters life. Being in part macroscopic and necessary, he is fated with the compulsion of necessity, and being in part kinetic and possible, he can make both himself and his environment other than they are. Being constituted of primary substances, he himself is a primary cause, and hence, makes a difference in things.

But only in the realm of ideas does real freedom exist, for only there does the kinetic microscopic principle operate freely in one's conscious experience. But what more than this could man desire, for forms, being physical, can affect the actual. The physical foundation of all forms renders ideas tremendously important. In fact, they are the only things which matter. All else is fate. At last we understand why one's philosophy of the first principles of science has had such an important effect on history.

But even with ideas man must be scientific to be happy. For ideas though physical must reckon with things as they are. All forms are not forms of the possible. It is for this reason that true ideas are necessary for a successful use of the possibles, and that man cannot live happily or effectively, in the world of art or of nature, without science.

But neither can man be truly happy without art. For the microscopic tendency of his nature must be given expression. The possibles have other uses than those involved in science in its discovery of the actual that is necessary; they have rights of their own. Precisely because of the peculiar rational character of his own nature man must be both scientist and artist in order to be himself. Einstein plays his violin.

References and Bibliography

1. H. S. Jennings. The Biological Foundations of Human Behaviour. Norton.
2. T. H. Morgan. The Theory of the Gene. Yale Press.
3. H. J. Muller. Artificial Transmutation of the Gene. Science, Vol. 66.
4. W. R. Coe. The Mechanics of Evolution. In Evolution of Earth and Man. Ed. by G. A. Baitsell. Yale Press.
5. H. Spemann. Organizers in Animal Development. Proc. of Royal Soc. London. Ser. B., 102.
6. H. Spemann. Naturwissenschaften, *32*, pp. 1–33 (1919).
7. O. Mangold. Das Determinationsproblem. Ergebnisse der Biologie. Bd III. Springer.
8. C. S. Sherrington. The Integrative Action of the Nervous System. Yale Press.
9. L. J. Henderson. The Order of Nature. Harvard Press.
10. L. J. Henderson. The Fitness of the Environment. Macmillan.
11. C. M. Child. The Origin and Development of the Nervous System. University of Chicago Press.
12. C. M. Child. Physiological Dominance and Physiological Isolation, etc. Roux Archiv. 117.

13. C. K. Davenport. The Relation of Neural Integration to General Physico-Chemical Organization, etc. Thesis. Deposited in Yale University Library.
14. W. P. Montague. A Materialistic Theory of Emergent Evolution. Holt.
15. F. H. Pike. The Driving Force in Organic Evolution. Ecology 10: 167ff.
16. F. H. Pike and others. Archiv. of Neur. and Psychiatry, Vol. 23, pp. 847–868.
17. H. S. Burr. Transplantation of Olfactory Placode in Amblystoma. Jour. Comp. Neur. 37: 455ff.
18. G. E. Coghill. Anatomy and The Problem of Behaviour. Macmillan.
19. C. J. Herrick. The Brains of Rats and Men. Univ. of Chicago Press.
20. K. S. Lashley. Brain Mechanisms and Intelligence. Univ. of Chicago Press.
21. J. Huxley. Essays of a Biologist. Chatto and Windus.
22. B. Russell. Introduction to Mathematical Philosophy. George Allen & Unwin.
23. Whitehead and Russell. Principia Mathematica. Cambridge Press.
24. Hilbert and Ackermann. Grundzüge der Theoretischen Logik. Springer.
25. H. M. Sheffer. Notational Relativity. Proc. 6th Int. Congress of Philosophy, Cambridge, Mass., 1926.
26. J. Royce. In Encyclopaedia of the Philosophical Sciences, 93ff. Macmillan.
26.[1] C. I. Lewis. A Survey of Symbolic Logic. Univ. of California.
27. L. Wittgenstein. Tractatus Logico-Philosophicus. Kegan Paul.
28. W. E. Johnson. Logic. Cambridge Press.
29. P. Weiss. The Nature of Systems. Open Court.
30. F. Enriques. Historic Development of Logic. Holt.
30.[1] A. Fraenkel. Einleitung in die Mengenlehre. Springer.
31. E. von Domarus. Das Denken und seine krankhaften Störungen. Kabitzsch.
32. D. Hume. Treatise on Human Nature. Longmans Green.
33. C. I. Lewis. Mind and The World Order. Scribners.
33.[1] E. Kant. Critique of Pure Reason. Trans. by N. Kemp Smith. Macmillan.
34. F. H. Bradley. Appearance and Reality. George Allen & Unwin.
35. G. E. Moore. External and Internal Relations. In Philosophical Studies. Kegan Paul.
36. W. James. Pragmatism. Longmans Green.
37. J. Dewey. The Quest for Certainty. Norton.
38. J. H. Jeans. Astronomy and Cosmogony. P. 351. Cambridge Press.
39. T. A. Cook. The Curves of Life. Constable.
40. J. Hambridge. Dynamic Symmetry. Yale Press.

CHAPTER VI

The Foundations of Experience and Knowledge

If the physical theory of nature is valid, consciousness must exist. This is not the usual supposition. It is customary to identify the kinetic atomic theory with materialism in metaphysics and extreme behaviorism in psychology. The falsity of this identification can be established by an appeal to purely objective evidence.

OBSERVED NATURE

The only theory of the first principles of science which is compatible with the denial of consciousness is one which entails no distinction between nature as conceived by scientific theory and nature as actually observed. A consideration of certain traditional theories will make this clear. The reader will recall how Greek inorganic science was led to conceive of nature as a system of unobservable geometrical and logical forms, and how certain of our contemporary physicists suggest a similar theory at the present time. Such a conception gives rise to a difficulty, as Plato saw. If nature is a system of ideal forms which only reason can grasp, why does the observed world of sensation exist? If one answers this question by saying that the world of forms, known by reason and defined by scientific theory, is the real world, and that the world of sensation, which we observe, is an appearance, then a conscious subject must exist as an additional natural factor with which the real world of forms combines to give rise to the sensible world of appearances. Only in a functional theory of nature, which, to use the words of Professor

Whitehead, avoids "a bifurcation of nature" [8] by regarding the colors and sounds and wets and drys of perceived nature as factors in the very entities of scientific theory, may avoid the admission of consciousness.

But the physical theory of nature does not do this. Although the extensive stuff and motion of observed nature are properties of the entities of the physical theory, there are other inescapable factors which are not. I refer to such obvious facts as the fragrance of the rose, the noise of the tolling bell, and the rich colors of the autumn foliage. Certainly neither the atoms nor electro-magnetic waves of scientific theory are fragrant, or noisy, or yellowish brown. In fact, the moment Galilei and Newton and the other founders of modern physics rejected the Aristotelian doctrine of opposites, for the thesis that rest is a form of motion, and cold a certain form of the molecular motion which is heat, etc., they doomed all who accept modern physics to the distinction between the world known by reason and the world known by sensation. Thus, although the physical theory of nature admits an identity between the world of theory and the world of sensation, this identity is not complete. A part of observed nature cannot be ascribed to the entities of physical theory, as they are defined by the physicist, and hence, assures us of the existence of consciousness to account for its presence in terms of the relation between the physical atoms and fields of scientific theory and the knowing subject.

It becomes evident that no psychological theory is more incompatible with the physicist's theory of first principles, than one which denies the existence of consciousness. For without a conscious factor in nature, the physical theory fails utterly before such inescapable facts as colors and sounds.

Consider the color of this page. Certainly the electrons which constitute the page are not white, nor is whiteness an attribute of the appropriate electro-magnetic wave which the physicist associates with the perceived color.

The same is true of any other secondary or tertiary quality in nature. Certainly a theory which does not account for the presence of such inescapable facts is incomplete or false. Yet, if nothing exists but the atoms and fields of the physical theory, as they are defined by the physicist, whites and blues and pains and pleasures would not be present. For no possible combination of colourless atoms or fields can ever give rise to a blue. By no process of logical gymnastics can one deduce a perceived color or sound or pain, or any other secondary or tertiary quality, from the entities of the macroscopic atomic theory as we have conceived them up to this point. To be sure, one must admit the appropriate atomic organization and wave formation which accompanies these qualities, but these physical and formal factors are not the qualities themselves, for they cannot be perceived, whereas the blue is seen. Up to this point we have accounted for only the physical, formal, and kinetic character of observed nature, and not for its colours and sounds and odours and pains and pleasures.

A great advance toward an adequate theory of secondary and tertiary qualities has been made by Professor Whitehead. He has taken advantage of recent advances in logic, making use of the notion of many-termed relations which are irreducible to two-termed relations of the traditional Aristotelian type. Thus the relation of the white to this page is not the two-termed relation of an attribute to a substance. Nor does it follow as Berkeley argues, because the white is not an attribute of the book, that it must be an attribute of the mind. All such arguments presuppose that the relation of a secondary quality to nature is a two-termed relation. Neither logic nor experience justifies such an assumption. The whiteness of this page is determined not merely by the physico-chemical character of the page, but also by the constitution of my eye, the nature of the intervening electro-magnetic medium, the lamp above my head, and many other factors. In short the relation of the whiteness of this page to nature is a many-termed relationship. But in the case of the physical theory of nature this is not all that is involved. The secondary or tertiary quality is complex in an intrinsic as well as a relational sense. For no mere increase in physical relata will give one an experienced blue. Although the atoms and their fields, and the many-termed relation joining their many compounds, determines that nature must exhibit a color rather than a noise on this book now, and having conditioned a color, prescribe that it is white, rather than red or green or blue, these purely physical and formal conditions do not give the experienced quality itself. For no mere combination of colourless atoms and fields can give a perceived color.[1] In short, if nature is nothing but the physical and formal properties of the atoms of the macroscopic atomic theory then the existence of colors and sounds is a mystery. Unless there is a subjective, conscious, psychical factor to combine with the physical and the formal in its many-termed relatedness to produce the secondary and tertiary qualities of the world of sensation, the physical theory of nature is condemned by the most obvious and inescapable of facts.

It is to be noted that the argument for the existence of the subjective and the psychical, which we have used up to this point, makes no appeal to human consciousness. It rests solely upon the consideration, which no philosophy with a pretense to the scientific can deny, that there are colors and sounds, and that these factors of objective fact are such that no amount of combination or deduction can ever derive them from only the physical and formal properties of the atomic entities which we have found it necessary to accept.

We know, therefore, that our theory of first principles cannot be complete until an addition is made to our organic atomic philosophy. In short, we must determine the precise nature of the psychical and indicate how it combines with the physical and formal principles of our theory to produce observed nature with its obvious colors and sounds as well as stuff and motion. In other words, the task which we now face is to reconcile the obvious presence of colors and sounds and pains and pleasures with the equally obvious extensive facts of stuff and change. Since the two latter facts find their basis in the macroscopic atomic theory, as we have previously defined it, this task resolves itself into a determination of the relation between the psychical factor in nature and the physical and formal properties of the atoms and their fields of our theory.

THE NATURE AND STATUS OF THE PSYCHICAL

The dual character of our knowledge of man enables us to solve this problem. Up to this point we have considered the atoms of the macroscopic atomic theory in the objective sense, as they stand in the relation of otherness to the knowing subject. But when we find that man *is* these entities in one of their many equilibria, another way of knowing them is available. For the knower is a man. Hence, when one senses what it is to be oneself, the atoms of our theory are joined to the knowing subject by the relation of identity; one knows the atoms which constitute oneself and nature by being immediately aware of what it is to *be* them. Now, I am conscious. Hence they must be also. Thus we discover that the subjective and psychical factor, which the presence of colors and sounds reveals, is an inherent property of the atoms of our theory. Man has a subjective character and is conscious, as he is rational and physical, because the ultimate atomic entities of which everything is constituted have psychical as well as physical and formal properties. Man is conscious because he *is* the entities of the macroscopic atomic theory in one of their many particularizations, which the pluralistic principle of this theory necessitates, and these atoms are inherently conscious. And observed nature is more than physical and formal nature, and is in part constituted by the perceiving subject, because the ultimate entities which constitute both it and its part, the observer, combine psychical with physical and formal properties in its synthesis.

Hence, because colors and sounds and pains and pleasures are in part psychical, it by no means follows that they are illusions or mere appearances, for even the psychical, which is a necessary factor in their existence, is as ultimate and irreducible and essential a property of atomic nature, as the physical and the formal.

The difference between the purely physical and formal world of physical and mathematical theory, and the observed world of stuff and change and form and colors and sounds and pains and pleasures, is not that the former is real and the latter an appearance. For, on the one hand, stuff and form and motion are as perceptible as colors and sounds and pains; and on the other, the dependence of secondary and tertiary qualities on a psychical factor does not make them a mere appearance, for the psychical is an attribute of ultimate reality possessing exactly as good a standing as its physical and formal properties. Moreover, as we shall note later, the determinate character which the psychical exhibits is conditioned by the physical and the formal. In fact, precisely because the atoms are psychical as well as physical and formal in character, they cannot compound and move to produce a world of matter and form without there being an experienced world of matter and form.

Thus, even though all men may die, there must be an observed and experienced world. For neither observed nature nor man can be what they are unless the ultimate atomic entities are psychical as well as physical and formal, and they cannot be this without an experienced world existing. It follows that there must be consciousness in the macroscopic compound nature which the macroscopic unifying principle of our theory introduces, as well as in the many locally-focused organic systems such as man, which the pluralistic principle of our theory necessitates. The theological significance of this will be noted later.

It appears that the realists are right when they say that colors and sounds are as ultimate as anything else; and that the idealists are correct when they assert that there is no nature apart from experienced nature. But this does not prove either that consciousness does not exist, or that all is mind. Instead, it means merely that minds as well as the objects of their awareness are constituted by one common system of atomic entities which possess physical, formal, and subjective psychical properties.

But to say that the entities which constitute both man and nature are conscious is not enough. It is also necessary to specify precisely what consciousness, as opposed to its content, is. Failure to do this has led to many misconceptions which have brought previous theories of this type into disrepute.

How can we determine the specific character of the conscious or psychical? It happens that there is a straightforward objective method. No speculations are necessary. Our analysis has revealed that the psychical exhibits itself not merely in the mind of man, but also objectively in observed nature. For it combines with the physical and formal conditions of observed nature to give colors and sounds and tertiary qualities. Otherwise there would be no difference between the world of physical theory and the world of observation. Hence, we have but to determine what the remainder in observed nature is, which the physical and formal does not produce, to discover the precise contribution and nature of the psychical.

The psychical is that in observed nature which the physical and formal properties of the entities of the macroscopic atomic theory do not condition. We have noted that this exhibits itself in the so-called secondary and tertiary qualities. But to assert this is not to maintain that they are completely psychical and subjective. In fact, we know that this is not the case. For observed nature exhibits stuff and change, as well as colors and sounds, and the former factors necessitate the macroscopic and microscopic atomic entities and their fields. Moreover, experimental physics has demonstrated that compounds of the atoms and their fields are uniquely associated with, and determine, the specific location and character of each experienced secondary or tertiary quality. Thus it is the physical and formal system of atomic structures and the intervening media, which prescribe whether nature exhibits a color or a sound or an odor in a given location, and if a color, whether it is red or green or blue. In short, modern science has shown that the physical and formal properties of the physical theory of nature prescribe the appearance, disappearance, and determinate character of any given sense quality. For example, the *determinate* character of the color of this page is as completely conditioned by the physical and formal properties of the systems of atoms and fields of the macroscopic atomic theory, as is the page's chemical constitution.

But if the psychical does not condition the determinate character and spatial or temporal extension and location of a given sense quality, then what does it condition? Certainly something is contributed; otherwise as we have demonstrated, there would be no difference between the world of physical and mathematical theory and the world of observation. Obviously, the psychical conditions that which remains, which is the bare indeterminate experienced quality that is made determinate by the physical and formal principles. Thus, by subtracting from observed nature, that which remains after the physical and formal properties of the atoms of our theory have made their full contribution, we discover the specific nature of the psychical. It is bare indeterminate experienced quality.

Note this definition with care: By an experienced quality we mean one that is immediately given, one that is not known by reason. And by experienced quality in its bareness and indeterminateness we mean immediately given quality or observed nature, abstracted from that which makes one of its instances or parts different from another.

To use the words of the English logician, W. E. Johnson, it is the determinable [2] of all determinables.

The correctness of this definition of the psychical can be demonstrated by an entirely independent analysis of observed nature. Consider hydrogen, oxygen, and water as they actually exhibit themselves in sensation. Their observed attributes have this peculiar character: the observed properties of one cannot be determined by an examination of the observed properties of the other. But this is not true of their physico-chemical or atomic character. From the combination of atoms of oxygen with atoms of hydrogen in proper proportion it follows necessarily that the product will be the atomic system which is water. This proves conclusively that there is a difference existing between hydrogen, oxygen, and water as conceived by physical theory, and these substances as actually observed. For the relations which join them in the two cases are not identical. We may summarize this by saying that the hydrogen, oxygen, and water known by reason in chemistry reduce to a common denominator and are commensurable, whereas in sensation they do not. In short, perception introduces an element of bare quality which must be experienced to be known, and which is so indeterminate that no deductions can be made from it. Thus we come again to the definition of the psychical as bare indeterminate experienced quality.

Idealism is correct in maintaining that perception contributes to the constitution of that which is perceived, but wrong in ascribing any of the determinateness of the perceived, such as form, or the secondary or tertiary qualities in their specificity, to perception. It errs also in certain of its many types, when it maintains that the subject or self is an ultimate irreducible entity. Both the knowing subject and known object are complex factors constituted of a common set of atomic entities which possess physical, formal, and psychical properties. It is absolutely essential that one distinguishes between the psychical and the psychological. The psychical is the psychological with the formal or logical, which makes it determinate, removed.

Some may wonder why the physicist has not discovered the psychical character of the atomic entities. But the reason for this is clear. Since the psychical is bare *experienced* quality it can only be known immediately. Thus the psychical character of an atom can be known only by being the atom in question. But the physicist, as physicist, only considers the atom in the relation of otherness to the knowing subject. Hence, his failure to find its psychical character is necessary.

Two objective determinations of the precise character of the psychical have been given. Both agree in defining it as bare indeterminate experienced quality. We noted that if atoms possess this property, it can only be known as a subjective fact when the object which it qualifies and the knowing subject are identical. These considerations enable us to deduce a consequence of our theory of the nature of the psychical, and to put it to another test. Because the barely psychical can only be known in itself, aside from its presence in objective observed nature when the knower and the entity or entities which the psychical qualifies are identical, and because man *is* all the atoms which the psychical qualifies in one of their many particularizations, it follows that the only system in which the barely psychical should exhibit itself to man in its own purity, unmixed with the physical and the formal, is himself. Hence, if our theory is true it follows that man will find a quality of being in himself which he finds immediately in no other system which he knows. Does our theory meet this test? The answer is in the affirmative. Man finds something in himself which he discovers immediately in no other creature or object. It is consciousness. Moreover, when one cuts off or abstracts from all physical and formal effects upon oneself, and turns back into the pure experience of one's own being, what does one find but the very definition of the psychical which we have given. Nothing remains but bare indeterminate experience.

It is of this that the mystic speaks. And it is precisely because it is experienced yet indeterminate, that he asserts it to be the most certain and intense of realities, yet cannot transmit it to others. Such is the nature of the psychical. It is incommensurable and untransmissible. And precisely because the observed and qualitative has this element of the ineffable and untransmissible in it, science avoids qualitative descriptions, for physical and mathematical theories. It is for the same reason that the behaviorist is quite right when he says that the concept of consciousness cannot be used in a scientific psychology, to account for the determinate character of mind. The psychical is made determinate by physical and formal principles; in itself it is not determinate, and it makes nothing else determinate. But this does not mean, as he goes on to assume, that consciousness does not exist. On the contrary it is present in every inductive science. For induction rests upon observation, and observation always gives the psychical along with the physical and formal, whether it be in psychology or geology. Experienced determinate mind is psychical, exactly as observed nature is psychical; no less, no more. In both cases one must look to physical and formal causes for the determinate character of one's subject-matter.

We find ourselves with a meaning for the distinction between consciousness and the content of consciousness. Consciousness, as opposed to its content, is bare indeterminate experienced quality. Its content is the determinateness added by the physical and formal properties of atomic nature. With this distinction clearly grasped the obvious experienced difference between man and other natural objects becomes intelligible. Failure to specify the precise nature of the purely psychical has led to the fallacious reading of the whole content of human consciousness back into the electron. Obviously, this is a ridiculous theory. But our doctrine that all the atoms of the macroscopic atomic theory are inherently conscious means no such thing, as the reader will immediately recognize if he keeps our definition of the psychical in mind. To say that they are psychical or conscious means merely that they possess bare indeterminate experienced quality in addition to their physical and formal properties. And precisely because of these physical and formal properties, this bare indeterminate experienced quality will take on determinateness; in other words, the atoms will have a content of consciousness as well as bare consciousness. But the physical and formal properties of an electron or of the macroscopic atom are different from those of man. Hence the conscious experience of an electron or the macroscopic atom will be radically different from that of man. In fact, precisely because the content of consciousness is determined completely by physical and formal conditions it follows necessarily that it will have little in common in an electron, a crystal, an amoeba, an anthropoid ape, a moron, or in an educated person. And for the same reason, since the physical and formal conditions of the content of their consciousness have so much in common, it follows that men will have a community of experience and feeling among themselves that does not occur in conjunction with other organic or inorganic systems.

It has been usual to identify the psychical with an entity called mind, or the soul, which is supposed to exist in addition to matter and body. All such identifications rest upon the failure to recognize what the precise nature of the psychical is. The moment we find it to be bare indeterminate experienced quality this is out of the question. In the first place, we are confronted with a quality rather than an entity. Secondly, since we have entities, in the atoms of our theory, to which this quality can be attributed, the principle of parsimony forbids the introduction of any more. But even if this objection were waived, the identification of bare experienced quality with an entity would be impossible. For to be an entity other than a physical substance a thing must have some determinations to distinguish it from that substance, and this is precisely what the psychical does not possess.

It is in this failure to recognize that the psychical, the

physical, and the formal are equally ultimate attributes of the atoms of which both nature and mind are constituted, that the gulf between Eastern and Western civilization has its basis. Generally speaking, the West has centered upon the physical and formal attributes of reality; whereas the East has concentrated more and more on the psychical; each identifying its object of attention with all that is real, and regarding the interests of the other as misguided or illusory. Precisely because the physical and the formal differentiate our experience, Western civilization has become increasingly complex and technical, leading on to an apparently endless series of problems which threaten to destroy human initiative; and because the psychical in its purity, is indeterminate, Eastern civilization has tended to move away from all differences to the bare experience called Nirvana in which the oneness of experience is grasped without its confusing specificities. It is to be noted that the Easterner who loses himself in Nirvana, and the negation of all that is specific, is as objective as the Westerner who masters physics to rear his steel mills, for bare indeterminate experienced quality is in objective nature as universally and unequivocally as the physical and formal which gives it its determinations. This enables us to appreciate why the Easterner tends to regard his pure experience to be the essential, and the physical and formal novelties which interest the Westerner, as but its incidental irrelevances.

But why identify man and his good with but one or two attributes of reality? The psychical, the formal, and the physical are equally real. Of these materials as they qualify atomic nature, observed nature and conscious men are constituted. One of this trinity of qualities is no better or no worse than another. Nor is there any opposition between them. The oppositions in reality are not, as traditional thought has supposed, between matter and mind, or matter and form; but between the macroscopic and the microscopic entities, all of which possess all of these three qualities.

It appears, therefore, that the issues of life must be

aligned on a new front. The truth is not to be attained, or proclaimed, by attacking the material or by praising the spiritual. Instead, the end of life is to be achieved by a proper adjustment of the possibles and the actual which enables all three basic attributes of man's nature to attain the fullest expression. When this is done the East will give more attention to the physical and the formal side of experience and the West will cultivate more of an appreciation of the sheer inexpressible experienced givenness in the determinateness of experienced nature. This it is which makes each observed occasion and each individual something unique and of intrinsic worth in itself. For it contains a character indefinable in terms of something else, which can only be grasped by possessing the experience itself.

At this point a great danger arises: the temptation to forget one's scientific and philosophical principles and turn the psychical into a cause of the determinateness of experience. When this happens, art and religion and science degenerate into sentimentalism. One of the most important tasks of philosophy is to clearly define the nature of the psychical, locate its place in the scheme of things, and keep it in that place. In this connection it is to be remembered that the determinate character of mind is as completely conditioned by physical and formal principles, as is the determinate character of a chemical element. All determinateness is physical and formal; the psychical contributes mere indeterminate experienced quality.

COMPLEX MAN

No account of man can divorce itself from physics, logic, or chemical and neural physiology. For these are the factors which give human experience its specific content.

Let us consider them in their bearing on the precise nature of human consciousness and the distinction between secondary and tertiary qualities. Our theory conceives of physical and formal nature as an equilibrium between the microscopic atomic entities of traditional physical theory and the macroscopic atom. Because of the pluralistic microscopic principle this equilibrium exhibits itself in many local compound systems; and because of the monistic macroscopic principle it gives rise to one grand macroscopic organization, which may be termed the order of nature as a whole. Thus compound nature is one large more static macroscopic order made up of many local dynamic organisms.

One of these local systems is man. It is because of the pluralistic principle of our theory, that he is aware of himself as here, and as something other than systems which are similar to him. In short, the individuality of man, as opposed to the individuality of other systems, or of nature as a whole, has its basis in the microscopic atomic principle of our theory. Moreover this principle is the source of our awareness of our body as a local object in a larger cosmos. And it is because conscious man is the product of a particularization of the conscious atomic entities of our theory, in which the pluralistic principle has come into the ascendency, that the consciousness which he possesses focuses around his body, and the character of its content is relative to that body.

But it follows from the principles of our theory that the human body is not a purely local system. Hence man's knowledge is not completely relative.

There are two senses in which we have found this to be true. In the first place, any living system is not a mere local compound of the chemical materials within its visible gross body. Instead, it is a dynamic equilibrium involving all the atoms of nature. It is because of this that one discovers all the entities of our theory to be conscious when one discovers oneself to be conscious. In the second place, even the gross local part of one's body, which the anatomist and student of gross physiology examines, exhibits the order of nature and the presence and partial character of other systems in its own organization. This was established in the previous chapter when we indicated how the unifying congesting influence of the macroscopic atom, operating through the order of nature as a whole, brings one local system into equilibrium and internal relations with the other systems of its environment. This element of necessary relatedness in man, enables his knowledge to escape thorough-going relativity, and makes science possible. Also man is not a purely local system. He is a localized equilibrium in equilibrium with other similar localized systems. This more complete equilibrium between local genetical and bodily materials, and the rest of nature exhibits itself in the nervous system, and particularly its central portion, the cortex of the brain.

Hence, if the entities which constitute man are conscious and the determinate character of consciousness is prescribed completely by physical and formal conditions, we would expect his consciousness to be most intimately correlated with the cortex of the brain and to include extensive objective nature as well as one's own body in its content. This happens to be the case. There are extroverts as well as introverts.

Moreover, since the organization and physical properties of the local phase of one's body are different from the structure and physical character of its environment phase, it follows that the experienced qualities of these physical and formal conditions must be different. This happens to be the case. Such a distinction has been discovered by students of the specific content of immediate experience. Those qualities associated with the environmental pole of the neural equilibrium are called secondary qualities, those connected with the more local or personal pole are called tertiary qualities or "bodily" sensations.

Of course, just as there is no rigid line separating the local part of one's body from its wider portion which is common to all systems and all nature, so there is no sharp division between the more personal tertiary properties such as pleasures and pains, and the more impersonal secondary qualities, such as colors and sounds. Hence no secondary quality is completely impersonal, and all possess feeling tone. It is because of the environmental foundations of one's own bodily being, and the internal relatedness, introduced by the macroscopic atom through the offices of the order of nature, which joins a sensitively organized local body to the physical systems which condition events in nature, that a complex of secondary qualities such as the halo of gold on the broken clouds of a western evening sky possess rich feeling tone and take on that melodramatic tinge which lifts one at times out of the commonplaces of the daily routine into the realm of ecstasy. In this fashion the secondary quality partakes of the intimacy and privacy of the tertiary.

And, on the other hand, because all "bodily" sensations, even those which are the most personal, owe their determinateness to physical conditions which are part and parcel of the one physical universe, even tertiary qualities are in nature. As Whitehead has suggested, the pain is in my tooth, and my tooth is in nature. Only a theory which recognizes that the specific character of sense awareness is completely conditioned by physical and formal factors, and proceeds to determine the specific organization of the physical which exhibits itself in nature and man, can give an intelligible and accurate account of the content of consciousness.

At last our theory of the nature of man nears completion. Like observed nature he is a synthesis of the psychical, physical, and formal properties of the metaphysical. In him the microscopic atoms and the macroscopic atom come into equilibrium, to produce not merely formed body with its polar opposition of the necessary and the possibles, but also to add bare indeterminate experienced quality to the determinateness given by the physical and the formal, thereby allowing metaphysical nature to constitute observed nature. His uniqueness centers in his peculiar and most complex physical organization. His body, because it represents a most delicately balanced adjustment, and because of its internal relation to the rest of nature, records and in part contains the macroscopic characteristic of nature in itself, thereby permitting an exceptionally rich and ever increasing experience.

In man, nature through the operation of its pluralistic localizing principle, and its monistic macroscopic principle has found a particularization of itself in which its trinity of ultimate properties, and its polar opposites are merged without giving one complete ascendency. Thus he has the rigidity of the stone coupled with the fluidity of the whirlpool, the static mold of necessity merged with the flow of possibility and creativity, the mortality of time stamped with the mark of eternity, the heritage of ignorance, together with the capacity for learning, and the power to pierce to the physical and formal depths and intricacies of things combined with a sense for aesthetic surfaces[3]. Synthesis of the physical, the formal, and the psychical, mixture of the flux of the temporal and the constancy of the eternal, carrying within himself the psychical sense of the Easterner, the overt physical emphasis and activity of the Modern, and the logical sense of the Scholastic and the Greek, he is able if he will to share and appreciate the experience of all peoples and all reality,—at once but an insignificant speck in a tremendous cosmos, and the created synthesis of every ultimate attribute of everything that is.

THE COMPLEXITY OF THE SECONDARY QUALITY

What is true of man the perceiver, is true of perceived nature. For nothing in nature, unless it be man, is more complex than such a commonplace as an observed color or sound. The supposition that they are simples in terms of which all else in science and philosophy is to be defined, is the basic error of most of modern epistemology. Consider a perceived blue. Before its determinateness, location, duration, and behavior is understood one must have a complicated theory of atomic structures, including man as well as nature, to which is added the bare indeterminate experienced quality which is rooted in the ultimate atoms themselves.

It is this which poetry tries to express. The poet hears the rippling brook and unlike the physicist he does not abstract the atoms and sound waves and forms of the running water from its ripple. Instead he tries to catch the whole of that elemental fact in its unity. If he is to succeed with us, he tells us that we must give of ourselves when we read his verse. In this he is correct, for the ripple is the psychical made determinate by the molecules and waves, and there can be no psychical factor in the object except as it is constituted by the experiencing subject. But this is not all. In addition, poetry at its best often suggests the existence of something grounded far beyond either the ripple or man, so far indeed that it must have its basis in the ultimate eternal nature of things. And in this also the artist is correct. For that which the poet or painter catches and the physicist misses is the bare indeterminate experienced quality which the psychical contributes, and this has its root neither in observed nature, nor in man, but in the ultimate atomic elements out of which both are constituted.

It is because art is dealing with experienced materials which, even in their simplest instances, are rooted in the physical, formal, and psychical properties of the metaphysical, that it is so difficult to achieve. Only a civilization which is reared on a philosophy that regards the psychical as ultimate can appreciate its importance, and only one with a philosophy which recognizes the *specific* character of experience to be determined completely by physical and formal principles can prevent its art from destroying itself in sentimentalism. For without the former belief, sensuous experience is regarded as a mere appearance, unworthy of serious attention, and without the latter all canons of excellence are destroyed. For sentimentalism is the doctrine that the *determinate* character of anything, whether it be a living organism, a mind, or a phantasy of the artistic imagination, is conditioned by the purely psychical. Since the psychical is indeterminate, ineffable, and inexpressible this removes all possibility of verifiability in science or intelligent criticism in art, and makes any messy outburst of repressed or uncontrolled emotions as genuine an artistic achievement as the creation of a Leonardo who has mastered the laws of his materials and forms. It was no accident that art and science were close together in the great Florentine.

But art is not exhausted in a synthesis of the physical and formal with the psychical. The grand symphony gives expression to the possibles of phantasy as well as to the actuality of imitative realism. At this point the polar principle of our physical organic atomic philosophy appears, and that aspect of art to which we referred in the last chapter exhibits itself. Because the physical and formal is constituted of a macroscopic one which imposes a general necessary macroscopic relatedness, and a kinetic microscopic many which introduces a series of changing forms of all possible kinds, art is at once the truest and most complete science of the actual and the fullest expression of the possibles. In it the understanding of the actual begins and ends, and the fruits of the imagination gain the recognition which is their due. At this point also our philosophy prepares the modern mind for the full appreciation of art. For it corrects the traditional supposition that the deliverances of imagination are less real than the findings of inductive science. The difference between the world of actuality with which perception deals and the world of possibility with which pure mathematics and the imaginative arts deal, is not that one is real and the other an appearance but that the former has its basis in the macroscopic atom, and the latter in the changing forms to which the moving microscopic particles give rise. It is because all nature and all experience is a synthesis of these opposites that physics is a mixture of necessity and contingency, art a combination of fact and fancy, and man an imaginative as well as an observing creature.

This synthesis of antitheses exhibits itself in the experience of men in another fashion. The sculptor before the block of marble would chip and mold it to the idea or form of the actual or possible which he has in mind. The intelligent leader of the social community would transform its present condition into some relatively more ideal state, the poet would create a personality nearer to the heart's desire. When this distinction between the actual and the possible exhibits itself in conduct one gets the moral life. It is because man and his life is constituted of possibilities as well as necessity that there are normative as well as empirical sciences. The distinction arises between that which is and that which may be. And it is because all nature and all experience is a synthesis of polar opposites that there is added to the possibilities of conduct, also the necessity of choice. Man not only may make a combination of the necessary and the possibles, but he must. Thus the moral as well as the aesthetic and logical character of man has its basis in the synthesis of the necessary and the possibles, to which the entities of the macroscopic atomic theory give rise. Because man is the macroscopic atom and contains within his being the effect of the one necessary form which it imposes upon everything, man is partially under the control of fate; because he is the microscopic particles with their inherent motion giving rise continuously to other worlds than this one, he is partially free to make himself as he lives; and because of the synthesis of these two parts of his nature which the congesting influence of the macroscopic atom introduces into every complex system, he must merge fate with freedom in a decision and take on a character of some kind. It is because the moving microscopic particles introduce a succession of possibles, and the macroscopic atom prevents any possible from remaining purely possible, that the moral life is unavoidable. It is for this reason, also, that morality degenerates into but another name for custom and prejudice, unless the imagination is cultivated, and one is guided by a philosophy which makes the possibles as actual and significant as the necessary.

GODS, SIMPLE AND COMPLEX

The foundations of experience have been considered as they exhibit themselves in observed nature, conscious man, the arts, and the moral life. They also merit consideration in themselves.

Recall what we know concerning them. They are atomic in character, and possess physical, formal, and psychical properties. This means that each atom has an immediate experience, the content of which is its own physical and formal self.

One has but to specify what this involves in the case of the macroscopic atom, to find oneself confronted with that collection of attributes which the Greeks identified with the divine being. In the first place, the macroscopic atom is a primary substance, with a determinate conscious experience. Secondly, were it not present neither nature nor man would exist; all would be flux. Hence it is the creator of all complex things. For creation, according to the physical theory, or any theory for that matter, which does not fall into the meaningless notion of making something out of nothing, means a rearrangement of the eternal microscopic particles into stable or partially stable systems. This, the macroscopic atom does.

Moreover, it accomplishes this result by merely being what it is. No mysterious or varying activity is involved. No change of its properties occurs. Its form is fixed, and its diameter is finite and small relatively to the tremendous number of microscopic atoms which it contains. Hence congestion, and change of direction of motion of the kinetic microscopic particles is necessary because of its mere presence. In this fashion man and complex nature come into existence. Hence, although, we use the expression, in a slightly different sense than did Aristotle, when he attributed it to the divine, the macroscopic atom is literally an "unmoved mover."[4]

Nor is the macroscopic atom even so slightly different from Aristotle's "unmoved mover," as one might suppose.

For it functions teleologically as the ideal toward which complex changing systems proceed, as well as the mechanical cause of their procedure. This is the case, as we indicated at the end of our chapter on the living organism, because the macroscopic atom has fixed properties and is an eternal entity. Hence its fixed form exists in the future as much as in the past. It appears before us in our relative time series, as an end to which we must adjust our conduct, as well as previous to us as the ground of what we do.

Those acquainted with the metaphysics of Aristotle, will note how our new organic atomic philosophy avoids the necessity which confronted him, of regarding God as a purely formal entity. His identification of contingency with matter drove him to this consequence, and the acceptance of the Platonic doctrine of forms which he had previously denied. For when astronomy revealed the existence of motions that are necessary—and it is to be remembered that Aristotle merely echoed the opinions of the empirical astronomers of his day on this point—he had no alternative but to make their causes purely formal. Because of our discovery that the opposition between contingency and necessity centers in the difference between the changing relations of the moving microscopic particles and the fixed form of the macroscopic atom, rather than in the opposition between matter and form, we are able to admit with the Aristotle of the first book of the Metaphysics that no form exists apart from a substance with material properties and at the same time account for necessary relatedness without going over to Platonism.

The macroscopic atom also enables us to escape another insoluble dilemma into which the functional theory drove Aristotle, and will drive anyone who follows it to its logical consequences. The reader will recall from the first chapter, how the fact of organization in biology and order in astronomical nature forced Aristotle to the conclusion that there is a formal as well as a material cause, and how the impossibility of accounting for the action of a pure form on a physical substance forced him to reject the atomic theory which conceives of matter as a substance for the doctrine that matter is a property along with form of a more fundamental type of substance, which is dynamic in character and hence is called the efficient cause. Thus at the beginning of his philosophy, Aristotle was emphatic upon two points. First, no form exists apart from a substance which has a material property as well. And second, not universals or forms, but many specific individual things are real; universals can be considered by themselves only in the mind of the thinker as abstractions from individual things. These two principles were perfectly compatible as long as Aristotle concerned himself with biology; there are many individual organisms and many local inorganic systems and in no case does a form or organization exist apart from their material constitution. But when he came to the order of nature as a whole, to the fact of relatedness or form between living things as well as within them, his philosophy broke down. He had to choose between denying all individuality to local living and inorganic things or admitting the existence of disembodied forms. It appears from Book Λ of his Metaphysics that he chose the latter alternative.

The necessity of a choice is clear. For if the order of nature which joins individual substances together is not to be a pure disembodied form, they, like the atoms in living organisms, must cease to be individual substances and turn into a mere material property which merges with the formal order of nature to constitute the only real individual—namely, complex nature as a whole. The result of this alternative is monism in metaphysics and pantheism in theology.

But Aristotle was a biologist and he had entirely too acute a sense of fact to deny the individuality of living things, combined with too good a sense of logical consistency to permit him to maintain both of his original premises. Hence he took the other alternative, retaining the individuality of local things and admitting the existence in macroscopic nature of the disembodied forms which he identified with God. There is no greater example in Western thought of intellectual honesty than Aristotle. Rather than ignore fact or deny logic he rejected one of the most fundamental premises in his philosophy. Such will be the fate of anyone who carries the functional theory of nature to its logical consequences. One must deny all ultimate individuality to everything except nature as a whole, or admit the existence of disembodied form. In either case an original tenet of Aristotelianism is denied.

Only by accepting the physical theory of nature as it is amended to include the macroscopic atom, can one be a consistent thorough-going Aristotelian in the distinctive and non-Platonic meaning of that term. For the pluralistic or atomic character of this theory provides a meaning for the individuality of men as well as of complex nature as a whole, and the physical nature of the macroscopic atom with its spherical form enables one to account for organization between living things as well as within them, without falling into the error of supposing that a disembodied form can change the physical character and motion of an embodied substance. It appears that the discovery of the macroscopic atom is an event of no mean significance in the metaphysical and theological world. For its being makes it possible to account for the necessary order and individuality of nature without denying an essential individuality to man, and to find a specific meaning for the divine without being forced to accept pantheism.

The physical character of the macroscopic atom has additional significance for theology. It means that God has a body. Otherwise it would be devoid of the power to change the direction of motion of the microscopic particles. This makes it evident that God is not omnipotent. For the microscopic particles possess energy also. Nevertheless, the macroscopic atom is more powerful than all the rest of primary matter combined. Otherwise it would be a complex substance rather than an atom, and the pressure of the microscopic atoms would burst it to pieces. Its atomic character renders the latter event impossible. This physical bodily character of the divine being, makes it a factor to be reckoned with in science and practical life. In fact, were God without a physical body, he would not exist. For the formal or psychical is impossible without the physical. Unlike the enfeebled, anaemic, purely spiritual creation of modern theological thought, the divine is not so impotent, and intangible, and completely divorced from observed physical nature, that one can trace the universe to its foundations without coming upon any evidence for its existence. Nothing can be more ridiculous than the rather prevalent notion that the divine can be as significant for life and conduct and the origin of things as its devotees claim, and at the same time fail to exhibit itself to exact science.

An attempt has been made to evade this conclusion by maintaining that science limits itself to that which is observed, and that the divine is ineffable and intangible, or in other words purely psychical, and hence, beyond the reach of science. Both parts of this contention are false. Science is dealing with the unobservable every day of its existence. It has done so throughout its past. And never was this more true than it is today. Its electrons, electromagnetic waves, its ψ functions, and four-dimensional continua are all unobservable. In fact, they are of such a nature that it is not merely practically but also theoretically impossible to observe them. But this does not mean that they are acts of faith or mere symbols. Science believes that they exist because their existence follows either necessarily or with a reasonable degree of probability from that which is observed. Hence, the statement that the divine is unobservable is no proof whatever that it cannot be known by science, providing, of course, that it exists. The statement that the divine is by nature purely ineffable, intangible, or psychical, is equally untenable. For the psychical is but the bare indeterminate experienced quality which is common to everything and hence cannot be the distinguishing characteristic of anything. It can be found in its purity as easily by concentrating upon bare indeterminateness in the experience of a dump-heap as by communing in a cathedral. Certainly this is hardly what one means by the religious experience. It must never be forgotten that the divine being can be distinguished from other objects only by its determinate character, and the determinate character of anything is physical and formal. Once this is recognized it becomes evident that the person who defines God as a purely spiritual or psychical being is really a thorough-going atheist. For a being cannot exist unless it has a determinate character, and this the purely psychical does not possess.

One must look to physical and formal properties for the attributes of divinity. Now, that which distinguishes the macroscopic atom from other simple and complex substances is its perfect spherical form. Since the formal is what we mean by the rational, it follows that the distinguishing attribute of God is rationality. Moreover, since the spherical form of the macroscopic atom is perfect, and this perfect form constitutes the content of its consciousness, it follows that this atom is the most perfectly and unequivocally clear-headed of all real objects.

This is in accord with the Greek tradition. Aristotle identified God with the system of forms. However, modern analysis of the foundations of mathematics and logic, and the nature of formal systems, has revealed that it is not necessary to state every form of a complicated system in order to completely define the system. Given a few atomic ideas or forms, the remainder follow necessarily. Our theory of first principles indicates that nature operates upon this basis. God is not, as Aristotle supposed, a musty museum of eternal forms. Instead the divine is the one embodied perfect atomic form, the sphere, which because of its relation of inclusion to the many rectilinear forms of microscopic atomic motion, generates, in the logical sense of that term, the complex forms of the world of imagination and observation. Some of these complex forms which are generated in this fashion are relations between moving microscopic atoms. Hence they come and go. We discover, therefore, that there are temporal forms as well as temporal objects. These relational forms constitute the world of the possibles. But because the form of the macroscopic atom is a property of an eternal substance rather than a changing relation between moving particles, the relatedness to which it gives rise in complex nature is necessary and eternal. This is the basis of the world of necessary actuality as opposed to the world of possibility.

Were nature constituted of nothing but the moving microscopic particles, experience would be a smudge of transition from one meaning that is not clear to another that is not clear, skepticism would be inevitable, and science impossible, since no necessary connections in experience would exist. It is the element of necessity and clarity which the macroscopic atom introduces into this smudge of vague feeling through the congestion and pressure of its eternal perfect physical form, that gives actual nature the element of necessary relatedness which must be present if science is to exist, and which enables human experience to take on what clarity and meaning it possesses. As the Greeks recognized, the divine reveals itself in man not in psychological experience, but in clear ideas. It was because of this that Socrates proposed to make the youth of the market place religious by forcing them to define their terms. Stated concisely, the essential and distinguishing attribute of God is intelligence.

It would appear, therefore, if to be religious is to be like the divine, that the one attribute above all others which one should prize is intelligence, as it exhibits itself in knowledge that is stated in terms of clear ideas or forms. For certainly, it is order in the realm of nature and clear consistent ideas in the realm of mind that the macroscopic atom is eternally engaged in producing.

But the macroscopic atom exhibits another attribute in its relation to man and complex nature. In our discussion of quantum and wave mechanics we noted that physical nature is a synthesis of opposites. This fact exhibited itself through every phase of natural and human experience. Space is a union of macroscopic constant uniformity and microscopic variable heterogeneity; atomic physics and optics reveal a continuous monistic wave medium merged in almost contradictory fashion, with a discontinuous pluralistic atomic principle; the living organism is an opposition between an externally conditioned organic factor and an internal pluralistic genetic factor; and undisciplined human consciousness is a mixture of sense and nonsense, truth and falsehood, and fact and fancy. In more general terms, rest is opposed to motion, constancy to variability, organization to entities, being to becoming, etc. In short, the macroscopic atom provides an antithesis for each thesis which the microscopic principle introduces. Hence, although it is on the side of order and calm being, it is also one source of the strife of nature, and the crises of human experience. But because of its spherical character, it brings its own attributes into conjunction with those of the microscopic particles to produce a synthesis of the antitheses. It is at once the ground of conflict and the source of its resolution. In this respect it is divine in the Hegelian sense.

Another attribute of the macroscopic atom is significant. Its union of perfect form with matter entails a complete break from traditional ideas. It has been customary to oppose matter to form. For example, Aristotle attributed the source of ignorance and the imperfections of the forms of sensible objects to the stubbornness of matter; forms are imperfect in concrete objects, he said, because matter prevents form from being perfectly actualized in individual things. In this way matter has been opposed to form, and body to reason. This vicious dichotomy has echoed down through the main channel of Western thought to divide the nature of man against itself. However, the divine object itself stands as an eternal reminder that matter and perfect form, or body and clear ideas are compatible. At last we have an idealism so inseparably grounded in a material realism that it cannot destroy itself in pious hopes. The macroscopic atom is perfectly formed, yet physical. Thus it serves both as the ground for existence and the ideal for all conduct. It is at once the source of intelligence, and its inspiration.

Nature and human experience have their antitheses, but instead of being between matter and form, they are between the formed complex matter produced by the microscopic, and the perfectly formed matter of the macroscopic. The good life is not something to be postponed until one frees oneself from supposedly material encumbrances, for this will never happen, nor is evil to be tolerated on the excuse that one has a body.

Again, it becomes evident that our organic atomic philosophy necessitates the alignment of the issues of life on a new front. In an intelligent use of the necessary and the possibles, life finds its difficulties and their resolution. The task of human existence, like that of the divine being, is an intelligent resolution of the antithesis between necessity and possibility. The physicists and mathematicians who pursue the necessary forms of actual nature into philosophy to find the macroscopic order which gives to existence what consistency and rationality it possesses are on the side of the divine, whereas those who call the youth of the land aside from the clarification of thought which comes with the pursuit of scientific knowledge, to search around in their own psychical interiors for some glow of self-satisfaction that is supposed to save their souls, are the real devils in the community.

This may sound to the reader like very radical doctrine. But as a matter of fact, it is exceedingly old. For it was said long ago by Plato, when he taught that what is most divine and good is in the realm of ideas rather than in the realm of sensation. It was precisely this which he meant when he wrote that one must "correct those corrupted courses of the head" by studying astronomy if one would be divine, and climb the dialectical ladder by bringing athletics, dancing, arithmetic, geometry, harmonics, and astronomy into that synthesis which enables one to see the part in the light of the whole, if one would live the good life.

Fortunately there are not so many barbarians pounding at our doors as was the case in ancient times, and those religious leaders who promise a short-cut to heaven which would avoid the hard climb up the ladder of scientific knowledge and practice, are enjoying more and more of the disrepute which is their well-earned desert. Perhaps Western civilization is at last prepared by its system of universal education for a religion which would make men divine by converting them, with a scientific education that has become philosophical, into rational beings like unto God whose most essential and distinctive attribute is perfect form.

It is to be emphasized that names do not give meanings to things. Like any assignment of a symbol, the attribution of the term, *divine,* to the macroscopic atom is a mere convention. Not one attribute of this atom is changed by this act. It remains after this assignment precisely what it was before. Moreover, if such a use of terms is made a justification for much that passes today in the name of religion, it becomes nothing short of a source of positive error. The nature of anything is determined by its properties, not by its name.

On the other hand, some rather commonplace properties often give rise to rather unexpected attributes when they are made specific. For example, consider the precise physical character of the macroscopic atom. If it is analogous in physical nature to the microscopic particles, we must assume, as we did in our theory of the ether or compound electro-magnetic field, that it is not merely a perfectly spherical hollow atom but also an inner field, exactly as the electron is not merely its central charge but also a field extending out from that charge. This means that every local system is constituted of the field of the macroscopic atom as well as the central charges and fields of the local microscopic particles. When we consider these two manifestations of the macroscopic atom, we find that it exhibits those two attributes of the divine which tradition has termed transcendence and imminence. The spherical shell of the macroscopic atom is a tremendous object off at the edge of the whole physical universe. This is God in the awe-inspiring overwhelming transcendental sense. But the inner field of this atom is in each one of us. This is God in the imminent sense. In fact, the body of man is partially the body of God. If this be true, then, since the consciousness of man is but the consciousness of his constituent materials, the actual calm perfect conscious rational experience of God is literally in the foundation of our own conscious nature. In short, there is an element of the divine mind in each human mind. But the macroscopic atom is imminent in us in another sense. For it imposes its form on complex nature to produce the macroscopic uniformity and constancy that is broken in microscopic regions by the variable heterogeneous relatedness which the kinetic microscopic principle introduces. This macroscopic relatedness, which originates in the macroscopic atom, operates through the order of nature to condition the organization of man. Thus in this indirect sense the macroscopic atom is imminent in man. Our conception of the nature of man enlarges. For with the identification of the macroscopic atom with the divine being it follows necessarily that man is not a purely temporal being. There is an element of the eternal, even of the divine, in him also. This point is as significant for the theory of knowledge as for theology.

Considered in its full psychical and transcendental sense, the macroscopic atom is a majestic being. Like us it has the thrill of experience for it is conscious of the physical and formal character of its own nature. And what a touch of the aesthetic and the calm there is to its finished perfect eternal form. Indeed, there is an aspect of self-sufficiency, finality, clarity, and beauty about that geometrical form which we know as the sphere. Certainly, sloppy unaesthetic loose ends and an infinite regress of dependence upon something beyond itself is out of accord with such a being. To be like it would be to replace the hectic clatter and restlessness of modern American life with the beautiful calmness of the Buddha at Kamakura. Over against the God of motion and change of modern life would be placed the God of beauty and calmness of the East. To "progress" one would add poise. Certainly, if the macroscopic atom is God, then to the rational character of the divine, we must add the aesthetic. In it the artist as well as the mathematical physicist finds an ideal. Beside truth one must put beauty, and if the imposition of calmness and poise and intelligibility upon experience is a value, then also goodness. Like many unassuming quite simple human creatures, the macroscopic atom becomes a truly divine and inspiring being when, with the eye of reason, one makes an acquaintance with the richness of its character.

Such is the meaning for divinity which seems to follow necessarily if the macroscopic atom exists, and all atomic entities are physical, formal, and psychical in character.

But this does not exhaust the theological implications of our theory. For it appears that there is more than one god.

The reader will recall the source of the individuality of man. The finite size of the macroscopic atom forces the entire system of entities of the macroscopic atomic theory to particularize itself in an equilibrium. This equilibrium is of a two-fold character. In molar and microscopic regions, the pluralistic principle of the theory exhibits itself most emphatically. This gives rise to the discontinuities of nature, among which are to be found those systems called men. But there is also a unifying organizing monistic principle which exhibits itself most emphatically in the macroscopic region of nature. This gives rise to the fact of continuity, to the field character of physics, and constitutes that most large and complex individual, the order of nature as a whole.

Since the organization of complex nature as a whole is even more uniform and constant and integrated than that of man, there seems to be no alternative but to regard it as an individual system. And since the character of this system is determined by the same psychical, physical, and formal atoms which make up man, there is no recourse but to regard it as conscious. Hence, it seems quite appropriate to regard it as divine.

This becomes even more obvious if one considers its character. Like man it is complex in nature. But because the macroscopic uniformity and constancy is in the ascendency in it, inconsistencies and conflicts do not constitute the center of the content of its consciousness. They are, to use an expression from William James, off in the periphery of consciousness. Hence it is naturally and instinctively a rational creature. It has to go out of its way to be illogical. Furthermore, it never dies. For having the source of its individuality in the eternal macroscopic atom, rather than in the changing pluralistic principle, its organization persists. Regardless of the microscopic or molar changes which occur, the macroscopic atom insures that the organization of nature as a whole must remain. Man dies merely because the pluralistic microscopic kinetic principle is in the ascendency in his bodily organization. Certainly a being which is conscious, eternal, and instinctively rational may be termed a divine being.

Hence, there are two gods. One, the macroscopic atom is a simple perfect substance. The other, the macroscopic unity of nature as a whole is a complex substance.

Many of the inconsistencies in theological theory have arisen because of the failure to distinguish between these two divine objects. The macroscopic atom is a perfect being, but because it represents only one form it is not omniscient. The complex order of nature, on the other hand, contains practically all forms, hence, except for the local microscopic or molar contingent forms such as appear in man, it is omniscient. But for this very reason it is not perfectly calm and happy, since the conflicts and discords and unfulfilled desires which arise from the opposition between the actual and the possible are implicit in the peripheral background of its nature. Like man it can be ill-proportioned, illogical, and evil, but since this is an exception to its instinctive rational nature, it can only achieve this with effort, exactly as we become logical only with long practice, if at all. But because of this potentiality this divine being can know the conflicts and pains of men.

Thus, whereas the perfect god of the world of being would inspire man by exemplifying eternal calmness, finished beauty, and perfect rationality, the god of the world of complex existence must add to these attributes an appreciation of the humor and the pathos of human experience. For even a god, if he brings within his consciousness the incongruities that arise out of the opposition and synthesis of the actual and the possibles, must at times laugh, and at other times cry.

And conversely, just as this complex being contains the contradictions and incongruities which the microscopic principle introduces, as a submerged part of its nature, so man contains the eternal and necessary macroscopic relatedness of the order of nature as an implicit knowable, but not an explicitly known part of himself. In this, the importance of education, the knowledge of universals, any possibility of immortality, and the principles of valid induction have their basis.

KNOWLEDGE

This brings us to the major theme of modern professional philosophy, that most difficult of all scientific problems, the problem of knowledge.[13] And because we have used the what of knowledge to provide a key to the how, in other words, because we have placed inductive natural philosophy, and physics, and metaphysics, ahead of epistemology, we are able to offer at least a preliminary solution of it. But first we must appreciate what the problem is.

It arose when Galilei and Newton and Dalton reared science upon the traditional physical theory of nature. This theory, because it was purely kinetic in its atomic character, necessitated the conclusion that no relations or connections in nature are necessary. The atoms can produce order as well as disorder. What is here today need not be here tomorrow. This metaphysics was stated in psychological terms by Locke and Berkeley. But Hume [5] saw that if there are no necessary connections there can be no causality, and without causality, no scientific laws. Few modern thinkers except Whitehead,[8] and perhaps Kant,[6] grasped the full force of Hume's logic. It is not, as so many of our contemporary scientists and philosophers suppose, that one must talk about probability or pragmatism rather than certainty whenever one makes a scientific statement. On the contrary, it means that if all connections are purely temporary then science should not exist. The consequence of Hume's analysis is neither probable nor pragmatic science but no science.

For there is no meaning to probability or pragmatism in the abstract. Probability is meaningless apart from a given specific set of conditions, and if these conditions do not hold for more than a moment there is no means of determining the probability of a theory formulated in one decade and tested in the next. Unless the conditions which define probability hold independent of time, in short, unless there is some necessary relatedness that can be counted on, even probable scientific theories are impossible. The same is true of pragmatism and its vague word "workability." Consider such a commonplace scientific operation as the measurement of an astronomical distance, or even more to the point, the joining of a measured value of such a distance determined in one century with another measured value made in another century to test the "workability" of a certain theory. Now, what does this workability presuppose? As Whitehead has emphasized,[5] and as we indicated in our chapter on the theory of relativity, at least general macroscopic metrical uniformity and constancy persisting through centuries is involved. If no relations in nature are necessary it is difficult to understand how this can be the case. In short, if the kinetic atomic theory in its traditional form is the whole truth, science should not exist.

But science does exist. Kant saw the point of Hume's analysis when he asked how mathematics is possible. Since he was a devout follower of Newton, and the mathematical aspect of Newton's mechanics was what most impressed men at the time, this was the same as asking why science exists.

The obvious answer is that the traditional kinetic atomic theory, which implies such a consequence must be either a false theory of the first principles of science, or only part of the truth. The philosophy of Whitehead, which replaces the physical theory of nature with a modern version of the functional theory rests upon the first alternative, the philosophy outlined in this book, upon the second.

Although Kant saw the problem, he pursued the wrong method in attempting to solve it.[6] Instead of returning to re-examine the premises which gave rise to the difficulty to reject or amend them in the light of a new review of the factual foundations of scientific knowledge, he mistook Hume's conclusion, which was but the consequent of a hypothetical proposition, for a true categorical proposition. Once having assumed that the physical foundations of experience give rise to no necessary connections, he had no alternative but to locate the necessary forms of experience in the mind. Science is possible, he said, because the mind constitutes the formal character of experience and the mind can think only in terms of certain necessary forms. This divorced philosophy from science by shifting the interest from physical and metaphysical to unverifiable epistemological issues, and left modern philosophical thought poverty-stricken before the existence of the possibles.

Kant's proposed solution received a fatal blow with the discovery of many geometrical and logical systems. This gave the lie to the thesis that the formal character of experience is constituted by the mind, which can think only in one formal way. C. I. Lewis has attempted to salvage what remains of Kant's theory by maintaining that the creative ego constitutes experience pragmatically rather than necessarily.[7] His work is important because it reveals that one who faces the discoveries of modern mathematics and logic can not be a Kantian without being a pragmatist. But pragmatism, as we have indicated, leaves us with Hume's difficulty, it does not get us out of it. Moreover, there seems to be little point in attempting to save a mortally wounded Kant, when the procedure which gave rise to his theory was faulty in the first place.

Instead of taking Hume's conclusion for granted, and then attempting to explain the existence of science with a series of epistemological hypotheses, we must re-examine the empirical foundations of his premises. This calls for a thorough-going review of the extensive facts of inductive natural philosophy and the technical evidence and theories of modern and contemporary science with special reference to one's theory of first principles.

Such a review and analysis has been undertaken in this book. At the very outset of science in the Greek world we found a most relevant and neglected fact. Observed nature gave them the notion of eternity and caused them to regard it as more primary than the concept of time. The first factor to impress Thales was the extensive fact of stuff. This gave Parmenides the principle of being, which means that the real does not change its properties. Certainly if anything does not change its properties it is eternal. The more we examine Greek science, the more this fact stands out. Even Aristotle, who accepted the principle of becoming, could never escape from the constant eternal aspect of things.[12] To be sure the fact of change was noted also, but it was but one temporal factor alongside of, or dependent upon, another eternal factor. Not until modern times did science take the concept of time seriously. And even in modern physics it seems to have no sound justification as an ultimate or primary concept. For traditional modern science accepted the kinetic atomic theory, and the logical consequence of this theory is to define time in terms of motion, rather than to define motion in terms of space and time.

But because of the problem of motion which the Greeks and moderns solved by identifying the required referent with the spatial characteristics of nature, space was taken as a primary concept. Hence when Galilei discovered time to be important, it was natural to regard it as a primary concept also. Thus a theory which must define time in terms of motion fell into the circular fallacy by defining motion in terms of absolute space and time. Once time is taken as a primary concept in terms of which all else is defined, in short once everything is regarded as a purely temporal system of relations, no necessary connections can exist and Hume's conclusion is inevitable.

But with the theory of relativity the doctrine of the primacy of time is given a death-blow.[12] Instead of absolute space and time being used to define motion, absolute light propagation is used to define time, thereby revealing that it is relative to different physical frames of reference. And in the general theory even space-time is found to vary with and hence be secondary to matter and its distribution or motion. Thus we are brought back again in physics to the primacy of matter and motion, exactly as the extensive facts of stuff and change force us to it in philosophy.

And at this point, the problem of motion reappears. For with the rejection of absolute time went absolute space. The question immediately arises: What is the referent for atomicity and motion. The evidence and considerations which force us to introduce the macroscopic atom have been given.

Once the traditional theory of first principles is supplemented with the addition of this atom, the problem of knowledge is resolved. The necessary relatedness in nature which must be present if science, as it is actually carried on, is to exist, has its basis in this atom. The varying relatedness which makes ignorance and error and uncertainty possible is rooted in the kinetic character of the microscopic particles. Thus the presence of science is accounted for in terms of principles which science discovers, without gaining such thorough-going internal relatedness that error and ignorance should be impossible. All relatedness and connections in nature are not purely temporal, for the structure of complex nature is a mixture of the changing relatedness which the kinetic principle introduces and the constant eternal form which the macroscopic atom imposes.

Moreover, the necessary connections and constancies must exhibit themselves in the general macroscopic characteristics of nature. Hence the Greek pre-Socratic philosophers were quite right when they claimed that the extensive characteristics of nature gave them the principle of being. Observation gives us propositions which are universal in an eternal as well as an extensive spatial sense. Moreover, since every local system, including man is a synthesis of the temporary relatedness of the microscopic principle and the necessary relatedness and being of the macroscopic atom, it follows that man is not a completely temporal creature. Being partially eternal in his literal physical constitution, valid induction is possible and the existence and validity of science is justified. For not only are singular, and hence universal propositions given in the observed macroscopic extensive characteristics of nature, but man is of such a character that he can know them.[12]

Thus we find ourselves at the end where we were at the beginning. In truly Greek fashion thought has run full circle. With the extensive facts of stuff and change we began, and with them, as conditioned by the kinetic and eternal character of the atoms of the macroscopic atomic theory we end. Not only have observed facts driven us to a new theory of the first principles of science, but this theory seems to have the fertility to account for their existence, our capacity to know them, and the scientific methods which are involved in the processes of knowledge. The modern anomaly of a theory of the first principles of science which denies the existence of science is no longer upon us. Science has become consistent with itself.

But in traversing the circle thought has enriched itself on the way. It finds the key to most of the difficulties of science and philosophy to be motion. And the secret about motion is simply this: it is. Other things are defined in terms of it, it is defined in terms of nothing else. But, although it is not defined in terms of other things, it, nevertheless, entails other things. In the first place, it entails that which moves. Here the only point to note is that qualities as qualities do not move, they simply are. Only matter moves. Other factors of experience such as qualities, or forms, or minds, derive what motion, if any, which they exhibit, by virtue of their qualification of, or determination by, physical substances. In the second place, motion entails *that,* relatively to which, a thing moves. Since there are no mathematical entities, and since there is no inertia of masses relatively to space but only an inertia of masses relatively to other masses, this referent must be a physical substance. Thus thought finds itself with a definition of matter. It is that which moves or the referent for that which moves.

Since there are many motions it follows that there are many physical objects. Hence nature must be constituted of many ultimate physical atoms moving relatively to a physical referent.

Besides motion, experience reveals continuity and order. In the course of its investigations science found it impossible to attribute the many evidences of continuous uniformity and structure to microscopic atomic motion alone. For each evidence of this inadequacy in its traditional theory of first principles, it introduced an absolute category. Thus in the course of history, space, time, gravitation, the ether, and irreducible biological organization appeared. In 1905 and in 1916 Einstein demonstrated that the first four of these absolutes are non-existent, as ultimate entities or forces, and that the facts which gave rise to them must be defined in terms of matter. Our analysis reveals the same to be true of biological organization. But the inescapable evidences of continuity and structure which give rise to these false absolutes cannot be defined in terms of the microscopic matter which moves, since that produces discontinuity and contingency of order. Hence, they must have their basis in the matter which is the referent for motion.

When we determine the type of physical referent which is necessary and sufficient to account for one of these evidences of continuity and constant structure, we find ourselves with a physical macroscopic atom which is capable of accounting for them all. The economy of thought which is attained is too great, as are the differences between the many false absolutes which it replaces, to attribute the macroscopic atom to a coincidence of human thought. Moreover, by accepting it in addition to the microscopic atoms of traditional physical theory, the difficulties of current science are resolved. We can admit that space-time and biological organization are to be defined in physical terms, as the general theory of relativity and physiological chemistry indicate, without finding ourselves with a theory of matter that is incapable of producing the uniformities and constancies which they involve. In short, we are able to reconcile the kinetic atomic character of nature, which modern science has unequivocally revealed, with its rational structure and metrical character, which Greek science discovered. There seems to be no alternative, therefore, but to conclude that the macroscopic atom exists, or, speaking more exactly, since only complex substances exist, that, like motion, it is.

Besides the fact of motion, and the microscopic particles and macroscopic atom, which it and the continuous structural facts of experience necessitate, we have also noted the rich colors and sounds and joys and sorrows of our experienced world and life, thereby discovering the specific nature of the purely psychical, and the inseparably united psychical, physical, and formal character of the metaphysical. Thus the real world, as revealed by science, when its analyses have merged into this philosophical synthesis, is the world of experience with which we began, with all its diversities and most melodramatic events made intelligible, while being robbed of the fearfulness which ignorance breeds. Hence the world with which we end is identical with, yet richer and more pleasant than, the one with which we began. Revealed as a synthesis of the psychical contribution of the subject and the physical and formal character of the object, both of which are constituted of the physical, formal, and psychical properties of the metaphysical in polar opposition and synthesis, one knows its physical and formal skeleton without its aesthetic flesh torn away.

Three principles seem to provide the key to the complexity, richness, and beauty of the scientific, aesthetic, and religious experience. They are the primacy of motion, the source of rationality and necessary order in the physical referent for motion, and the identification of the purely psychical with bare indeterminate experienced quality. When these principles are combined and carried to their logical consequences it is not with dead concepts but with living experience that we end. Reality as known by this scientific philosophy is aesthetic immediacy with its physical, formal, and psychical conditions made specific. It combines the realism of the empiricist with the formal idealism of the rationalist, the surface immediacy and appreciation of the artist with the depth and analysis of the scientist. If one were to pick one man in Western history who more than any other came nearer to the full richness of the truth, the choice would have to fall neither on Plato, nor Aristotle, nor even Democritos, but on the great Florentine, Leonardo da Vinci. For with him the physical and the formal were grasped without being torn from the vivid psychical immediacy in which both are imbedded. Aesthetic and religious experience and scientific knowledge are indissolubly wedded because the foundations of immediate experience and rational knowledge are one and the same. But of course, this is only true for those who use exact science to determine first principles, and having arrived at a theory which seems to be the most probable and fertile, modify their artistic, religious, and scientific conceptions and practices accordingly.

References and Bibliography

1. W. P. Montague. Belief Unbound. Yale Press.
2. W. E. Johnson. Logic. Vol. I. P. 174. Cambridge Press.
3. D. W. Prall. Aesthetic Judgment. Crowell.
4. Aristotle. Metaphysics. Trans. by W. D. Ross. Oxford Press.
5. D. Hume. Treatise on Human Nature. Longmans Green.
6. E. Kant. The Critique of Pure Reason. Trans. by N. Kemp Smith. Macmillan.
7. C. I. Lewis. Mind and The World Order. Scribners.
8. A. N. Whitehead. The Concept of Nature. Cambridge Press.
9. A. N. Whitehead. Science and The Modern World. Chs. X, XI. Macmillan.
10. A. N. Whitehead. Religion in the Making. Macmillan.
11. A. N. Whitehead. Process and Reality. Macmillan.
12. F. S. C. Northrop. The Relation Between Time and Eternity. Proc. 7th Int. Cong. of Phil. Oxford, 1930.
13. D. C. Macintosh. The Problem of Knowledge. Macmillan.
14. W. E. Hocking. The Meaning of God in Human Experience. Yale Press.
15. J. Y. Simpson. Nature: Cosmic, Human, and Divine. Yale Press.
16. C. D. Broad. Mind and Its Place in Nature. Harcourt, Brace.
17. B. Russell. The Analysis of Mind. George Allen and Unwin.
18. J. Ward. Naturalism and Agnosticism. Black.
19. S. Alexander. Space, Time and Deity. Macmillan.
20. J. M. E. McTaggart. Studies in Hegelian Cosmology. Cambridge Press.
21. J. M. E. McTaggart. The Nature of Existence. Cambridge Press.

INDEX

Printed in the United States
By Bookmasters